QB 280.5 .M87 2009
Murdin, Paul.
Full meridian of glory

MHCC WITHDRAWN

D0828522

Full Meridian *of* Glory

Perilous Adventures
〜 in the Competition to 〜
Measure the Earth

Full Meridian *of* Glory

Perilous Adventures
⊹ in the Competition to ⊹
Measure the Earth

PAUL MURDIN

Copernicus Books
An Imprint of Springer Science + Business Media

ISBN: 978-0-387-75533-5   e-ISBN: 978-0-387-75534-2
DOI: 10.1007/978-0-387-75534-2

© Springer Science + Business Media, LLC 2009
All rights reserved. No part of this publication may be reproduced, stored in a retrieval system,
or transmitted, in any form or by any means, electronic, mechanical, photocopying, recording,
or otherwise, without the prior written permission of the publisher.

Published in the United States by Copernicus Books,
an imprint of Springer Science+Business Media.

Copernicus Books
Springer Science + Business Media
233 Spring Street
New York, NY 10013
www.springer.com

Library of Congress Control Number:
2008939455

Manufactured in the United States of America.
Printed on acid-free paper

9 8 7 6 5 4 3 2 1

I have touched the highest point of all my
greatness;
And from that full meridian of my glory
I haste now to my setting.

William Shakespeare
King Henry VIII. Act iii. Sc. 2.

Contents

List of Figures

Introduction

In the spring of 1962 I was studying mathematics and physics at university in England and looking out for something to do in the summer. A notice appeared in the porter's lodge inviting applications to spend eight weeks at the Royal Greenwich Observatory (RGO) on an astronomy course. In the mornings there would be lectures on all the facets of astronomy carried out at the RGO, in the afternoons students would work on a research topic as assistants to the scientists there. The course was residential and students would stay in Herstmonceux Castle in Sussex, where the RGO was based. We would be offered the accommodation and paid a small allowance as well. I had been an amateur astronomer while at school so I would be paid for doing something I liked. This all sounded great! I applied, and was accepted on to the course.

I very much enjoyed this first experience of professional astronomy. My own research topic was to measure the positions and sizes of the spots that represented stars in some photographs of the globular cluster Omega Centauri. I combined my measurements with those that other people had measured on other photographs. I was able to see how the stars moved over time. I could distinguish stars in the cluster from stars in the line of sight in front of and behind the cluster. I could find stars whose brightness changed from photograph to photograph and even discover how some changed rhythmically like a regular heartbeat. I learned how to connect these measurements with the distance of the cluster and to determine its orbit in our Galaxy. It was mindblowing that so much could be learnt from pictures of spots.

It was this experience that made me determined to be a professional astronomer. At the end of the course each student was summoned, one by one, to the office of the Astronomer Royal, Sir Richard Woolley. I entered to see him flipping through what I realized must have been reports on what I had been doing. He raised his head and fixed me with his eyes, glaring over the half moons of his reading glasses. "So what's your story on Omega Centauri?" he barked. I stumblingly described what I had done, he asked me more questions about it, and we traded hypotheses and facts for twenty minutes. "What are you going to do when you go down from Oxford?" he asked. "I want to be an astronomer," I said, with new conviction. "Come and work here," he said. Somehow it didn't seem strange to be offered a job in this way. "OK," I said. "Thanks." My cool reply concealed the exultation within me.

This is the way that I came to work at the RGO in 1963. I started to work on problems that were akin to my student project, about the motions and the positions

of stars – so-called positional astronomy. Although I soon shifted my area of specialty towards astrophysics (the study of the internal constitution of the stars, galaxies and planets), I learned more about positional astronomy and its long history at
the RGO. Three hundred years before I joined it, the Royal Observatory had been
established at Greenwich for the very purpose of measuring star positions in order
to help the sailors of the British Royal Navy and the Merchant Navy navigate safely
across the sea. When I joined the RGO, although the observatory had been shifted
from Greenwich to Herstmonceux, it remained a primary task to continue this work
in modern form.

It was a matter of some pride to all of us who worked there that the RGO had
this long history and had established its eminence in this work on the world stage.
Of course, we all knew of similar observatories elsewhere that had carried out parallel studies, each making their own advances and their own contributions.
Foremost among these, but definitely in second place in our opinion, was the Paris
Observatory in France. It had been founded before the Royal Observatory in
England, in fact, and it had carried out a variety of work similar to that of the work
at Greenwich. There had been a certain rivalry between the two establishments,
which mirrored the traditional rivalry between the two countries, but both had
shifted their focus to modern astrophysics, while still retaining interest in the fundamental studies of positional astronomy.

There was one area in which the Royal Observatory at Greenwich appeared definitely to have been favored over the Paris Observatory, namely in the decision
(Chapter 7) to define the Greenwich Meridian as the "prime meridian" of the world,
from longitude would be measured, and the associated decision to use Greenwich
Mean Time as the world's basic time zone. This decision was founded, as I interpreted it then, on the pre-eminence of the scientific work at Greenwich to measure
time and position as a starting point for marine navigation. I took some pride, not
much earned by me, that I was a member of the scientific establishment descended
from the one that carried out this work.

It was some years later that I came to realize how the decision to choose
Greenwich as the prime meridian had been made and the outstanding scientific
work that the rival Paris Meridian had been based on. I was making repeated visits
to France and this brought me into closer contact with the country. As well as going
to France purely for pleasure, I had a job with the British National Space Centre,
which involved frequent visits to the European Space Agency, whose administrative headquarters are in Paris, so I often traveled there. I thus came to learn about
the structure and history of French science and the scientific work of the Paris
Observatory. It was astounding. I gained a new respect for the Paris Observatory
and was rather ashamed of my earlier limited view of the story of the development
of the science of navigation, which I had viewed as almost entirely confined to the
work of the Royal Observatory at Greenwich.

As part of my job, I had also become accustomed to think about problems of Big
Science because of my work with space experiments and large astronomical
telescopes. The sums of money are large, the numbers of people are high, the range
of skills is huge, and the management tasks are considerable. Scientists get involved

in such projects because they are targeted towards scientific goals that they think will be highly rewarding. Usually, because of the special talents that need to be assembled, the scientific teams are international. Additionally, the sums of money are so large that it helps if the costs are shared among several countries. This means that there is an international political dimension involved in such projects. These Big Science projects only succeed because the scientists work well together. They are, however, human beings and they do squabble sometimes.

I found all these features (which I had assumed were mainly modern) in the history of the work of the Paris Observatory to establish the Paris Meridian. The Paris Academy of Sciences had pioneered the concept of scientific *grands projets*. The community of scientists who worked on the Paris Meridian project for over two hundred years was large and international. They knew each other well – some were members of the same family, in one case of four generations. Like scientists everywhere they collaborated and formed alliances to achieve their goals; they also split into warring factions and squabbled over the ways to do so. They traveled to foreign countries, somehow transcending the national and political disputes, as scientists do now, their eyes fixed on ideas of accuracy, truth and objective, enduring values – save where the reception given to their own work is concerned, when some became blind to high ideals and descended into petty politics, small feuds and personal attacks. They organized themselves into effective working teams but also their projects ran over in time and budgets were overspent, just like now.

To establish the Paris Meridian, the scientists endured hardship in atrocious weather as they traveled in rugged, remote places (Chapters 2, 3, 5 and 6). They survived danger and gloried in amazing adventures during a time of turmoil in Europe, the Revolution in France and the Napoleonic wars between France and Spain. Some were arrested and imprisoned. Some were accused of witchcraft. Some of their associates lost their heads on the guillotine. Some died of disease. Some won honor and fame. One found irredeemable fault with his own work and became a depressive loner. One became the Head of State in France, albeit for no more than a few weeks. Some found long-lasting love or brief and dangerous flirtations in foreign countries. One scientist killed in self defence when attacked by a jealous lover and another was himself killed by a jealous lover. A third brought back a woman to France and then jilted her; destitute, she retired to a convent.

It is true that modern adventures in the pursuit of Big Science tend to be less dramatic. I have never had an observing run on a telescope that was as full of incident as Arago's (Chapter 6). But I have seen an astronomer on his way to a telescope on a bus that caught fire desparately searching for his luggage piled in the smoke on the roof of the bus, in order to save the papers that he would need to use to set the telescope. Another astronomer returning from his observing trip survived the incident in 1988 when a Boeing 737 of Aloha Airlines that was flying between islands in Hawaii lost its upper fuselage in flight. And telescopes are built on mountain tops which are always dangerous places to be with high cliffs, thin air, snow and icy weather.

The French scientists of the eighteenth and nineteenth centuries worked on practical problems of interest to the Government, the Church and the People (Chapters

2, 4 and 5). They also worked on the important intellectual problems of the time, including a problem of great interest to their fellow scientists all over the world, nothing less than the theory of universal gravitation (Chapter 3). The Paris Meridian was a construct that tested the theory of gravity, the same theory that drops apples from trees, shapes the Earth, holds the planets in their orbits and, indeed, decides the fate of the Universe.

French scientists succeeded in their intellectual work, while touching politics and the affairs of state (Chapters 5 and 6). Their endeavors have left their marks on the landscape, in art and in literature. However, in one regard the chance to leave the mark of their scientific work on everyday life slipped from their fingers. As a result of the international negotiation over 100 years ago, in a decision based on commercial and political values more than scientific ones, the Paris Meridian was not chosen as the prime meridian of the world (Chapter 7). The world might have talked of Paris Mean Time instead of Greenwich Mean Time, and we might have each measured our location on Earth with respect to the Paris Meridian, not the Greenwich Meridian. This did not happen.

In no way did this disappointment detract from the historic scientific importance of the French work. Anyone who visits the Paris and the Greenwich Meridians is making a pilgrimage of great significance, although in both London and Paris the practical significance of the meridian lines has faded with modern developments (Chapter 8). The line in the cobbled courtyard at Greenwich which marks the Prime Meridian memorializes the international agreement that organized time across the whole world and harmonized its longitude systems so that ships could sail more safely, the world acting as one body for the safety and convenience of its human population. The line in Paris is also marked on the ground, across the whole of Paris from one side of the Péripherique (inner orbital road) to the other. It takes the form of a series of over 100 brass disks (Chapters 10), ostensibly a memorial to the astronomer François Arago, who worked to make some of the scientific measurements that defined the Paris Meridian, but also implicitly a memorial to all who worked before, with and after him on that *grand projet*.

The Paris Meridian and this line of disks has recently achieved notoriety as a result of the popularity of Dan Brown's novel *The Da Vinci Code* (Chapter 9) but it should be famous as a testimonial to Arago and the other scientists who worked to define the Paris Meridian, both for its practical utility and for its scientific impact. However, I have to confess that it was reading the *The Da Vinci Code* as a holiday entertainment that rekindled my interest in the Paris Meridian and took it beyond idle curiosity. As a result, like many tourists, I set out one summer's day to follow the Paris Meridian across Paris (Chapter 10). As I walked the line, my eyes towards the ground, scanning in search of the next Arago disk, my imagination soared into the sky, inspired by the science and technology that lies behind the Meridian line and the dedication and brave adventures of people in search of the science of the Universe. I decided to tell this story of the Paris Meridian, complete, here in English for the first time.

The stories in my book are not fiction, they are history. The era that I describe was historical but the scientific work was both ageless and modern. The motives for the endeavors that I describe were neither religious nor conspiratorial, they were scientific. I see the line of the Arago Memorial on the Paris Meridian, not as part of a silly conspiratorial plot (as in Dan Brown's novel), but as an expression of the human spirit of scientific curiosity about the Universe. I intend this to be a work of inspiration, not of original historical research. I write about the Paris Meridian in the spirit of the Serbian proverb: "Mankind! be humble, for you are made from earth; Mankind! be noble, for you know the stars."

Institute of Astronomy, Cambridge,   Paul Murdin
April 2008

Chronology

1656	Huygens invents his pendulum clock
1661	King Louis XIV takes the throne of France
1665–66	Colbert founds the Académie des Sciences
1668–70	Picard measures the meridian from Malvoisine to Amiens
1669	Cassini I becomes the leader of the Paris Observatory
1671–72	Picard determines the longitude of Hven
1671–72	Jean Richer's expedition to Cayenne; experiments with the pendulum
1672	Paris Observatory completed
1672–82	Picard, Cassini I and La Hire I determine the positions of the coasts of France
1687	First Edition of Newton's *Principia*
1683	Cassini I and La Hire I measure the meridian from Montluçon to Sourdon
1702–13	War of the Spanish Succession
1712	Cassini II becomes the leader of the Paris Observatory
1715	Death of King Louis XIV, accession of his great grandson, Louis XV
1718	Cassini II, La Hire II and Maraldi I measure the meridian from Perpignan to Dunkerque
1725-28	Voltaire visits London
1728	Publication of the *Encyclopédie*
1734–37	Maupertuis, Le Monnier and Celsius measure the degree in Lapland
1735–44	La Condamine measures the degree in Peru
1733	Cassini II and Cassini III measure the East-west axis of France from Brest to Strasbourg
1739	Cassini III, Lacaille and Maraldi II measure the meridian across France
1742	Le Monnier completes the *meridiana* in St Sulpice
1754	Death of King Louis XV, accession of his grandson Louis XVI
1765	Cassini III appointed leader of the Paris Observatory (named director 1771)
1784	Cassini IV becomes Director of the Paris Observatory
1787	Roy, Cassini IV, Legendre and Méchain link the Paris Meridian to Greenwich
1789	Storming of the Bastille

1790	Cassini IV publishes the Cartes de France
1792–3	Overthrow and execution of King Louis XVI and foundation of the French Republic
1792–98	Delambre and Méchain measure the meridian from Dunkerque to Barcelona
1792–1800	The French Revolutionary Wars across Europe
1799	Napoleon Bonaparte becomes leader of France as First Consul (Emperor from 1804)
1806–08	Biot and Arago extend the meridian from Ibiza to Barcelona
1809	Arago returns to France after being imprisoned four times
1808–14	The Peninsular War
1814	Napoleon abdicates, is exiled to Elba
1815	Napoleon escapes from Elba, re-establishes power but is defeated and exiled this time to St Helena
1884	Washington Conference on the Prime Meridian passes over the choice of the Paris Meridian
1911	France adopts the Greenwich Meridian as the basis for longitude

Chapter 1
The *Incroyable Pique-nique* and the *Méridienne Verte*

Places can be admired for what they *are* but they can also be admired for what they *mean*. The White Cliffs of Dover on the north shore of the English Channel between Britain and France are a beautiful seascape, the white, massive, motionless, white chalk cliffs contrasting with the active gray-blue sea. The Cliffs are also a symbol of the island independence of Britain, the bulwark shore that historically kept out potential invaders. For Britons returning via ferries and liners from overseas, the White Cliffs may be the first sight of home after a long absence, providing a warm welcome home. Americans may react similarly to the New York skyline and the Statue of Liberty, South Africans to Table Mountain, and Australians to the Sydney Harbor and its Bridge.

It is not often that people react to an abstraction of a place. The Paris Meridian is the north-south line running through Paris. It is entirely theoretical and there is no one landscape of beauty to react to. Rather, it cuts through a collection of typical French landscapes, the good and the bad proportionately represented. The Paris Meridian runs through the French capital and a sample of its neighborhoods. It breaks out of the capital through the suburbs, some charming, some depressed. It runs north through both rural and industrial France to Dunkerque and the sandy dunes of the northern French coast. It runs south alongside the major autoroute, La Méridienne, and through the mountains of the Massif Central, the central block of elevated land in the middle of France. It then drops through the vineyards on the slopes of southern France. Beyond Perpignan, the southernmost city in France, it rises to the mountains of the Pyrenees and continues into Spain, past Barcelona and into the Balearic Islands.

Perhaps even some French people do not know exactly what the Paris Meridian is. However, they are proud of the magnificent achievements of the astronomers and geodesists who created it, and who, using it as a base, meticulously mapped France and measured the world. This pride in the theoretical was used in the argument two hundred years ago that the distance along the Paris Meridian from the North Pole to the Equator should be used as the fundamental international standard of length – the meter.

Standardization helps unite a political region. For example, In 221 BCE in the Eastern Zhou region of Asia the Emperor Qin (or Chi-in) unified seven of the warring states, subjugating rival states through ruthless centralization and imposing standard

P. Murdin, *Full Meridian of Glory*,
DOI: 10.1007/978-0-387-75534-2_1, © Springer Science+Business Media, LLC 2009

legal codes and bureaucratic procedures. His idea, soaked in the blood of his enemies, was that he would unify the country to which his name is given – China – by establishing standard forms of coinage, weights and measures. He also standardized the various spoken languages in his kingdom into written pictures drawn the same no matter what their spoken form so he could receive reports that he could read from his officials throughout the country. He even established a standard length for the axles of carriages so that it was possible to drive through archways and over bridges along roads from border to border. These measures made it possible to communicate and trade across the whole of the empire, knitting the country together.

Similarly, when the new government of France took command after the French Revolution late in the eighteenth century, it took on the problem of the development of a system of weights and measures to unify the French people across the entire country. Gallic logic and communard feeling suggested that standards of weights and measures should be not based on arbitrary diktats laid down from above, especially not based on such a transient thing as the size of a monarch's body. Standards of measurement should also not be considered matters of national authority (a stance which practically guarantees that the standards would not be internationally adopted). International standards, making it possible to develop both trade and communication between different people, should be based on natural quantities that belonged to no one and to everyone. This was the origin of the meter and the metric system of units. The French people have continued this historical direction for two hundred years and even now are pushing to improve trading standards across the European Union.

THE MERIDIAN also played a part in making what some would argue was the most significant discovery of science – the law of gravitation. Curiously this scientific law of remote celestial bodies was a turning point, not only for the abstract science of astronomy but also for science in general. Sir Isaac Newton's mathematical law of gravity changed our perception of the laws of nature. It made understandable in mathematical form rules of nature that otherwise seemed arbitary. It enabled scientists to predict the future, even the return of a comet after 74 years invisibility in the far reaches of the solar system. It set the standard to which scientists now aspire in developing the truths in their own science.

Newton's theory also altered our perception of ourselves and the Universe we live in. Since the law of gravitation applies equally to the planets and to objects on the surface of the Earth, such as a falling apple, it puts us into the Universe; we are *a part* of the way that it works, not *apart* from it. This scientific perspective found resonance in the political and social egalitarianism of the revolutionaries and the humanists.

The Paris Meridian played a role in changing scientists' minds about gravitation. Measurements along the meridian of the shape of the Earth were instrumental in showing that Newton's Theory of Gravity was convincingly correct. But for the French there is also a bitter sweetness about the Paris Meridian. First created in the seventeenth century, and an admired scientific work, its status fell at the end of the nineteenth century when it was passed over as the choice of the Prime Meridian of the world. Following France, many countries had developed maps based on meridians through their national observatory thus causing great confusion among sailors when they sailed from one map to another issued by a different authority. The development of international trade across the world created a demand for a unified, standardized system of coordinates of latitude and longitude. Latitude was easy; everyone agreed that it was measured from a zero-point at the Earth's equator but the zero-point of longitude – the "Prime" Meridian – was not as obvious. People asked should it be one of the national meridians or should it lie through some natural or artificial feature (such as the Great Pyramid in Egypt)?

In 1884 a conference was convened in Washington to decide the issue. The Paris Meridian was an obvious contender because of the work that had been put into its accurate definition and its key place in scientific development, and so was the Greenwich Meridian (for similar reasons). The logic for the definition of the meter was that a neutral meridian should be chosen. Just as the meter favored no individual person as the basis of a scale of length (the reach of a king's arm, for example, or the length of his foot), the choice of a neutral meridian as "Prime" would favor no particular nation. This logic was ignored, though, and for practical reasons the Greenwich Meridian was chosen as the Prime Meridian of the world.

The status of "Prime" Meridian was wrested from Paris and given to Greenwich. Some say this was by the logic of nineteenth century commerce and power, some say that the USA achieved its objectives by bringing the quarreling Europeans together and forcing them to a conclusion, and some say that the Prime Meridian was established at Greenwich by an anglo-phone conspiracy against France that was formed between Britain and the USA. Sometimes French, British and American people see contemporary events in a similar way.

Even if the Paris Meridian lost its status in these considerations of *realpolitik* (politics based on practical rather than logical or ideological considerations), it retains its status in the French psyche. Laid out on a map (Fig. 1), the meridian has a geometric cleanliness, running directly across the center of the hexagon of the entire country,[1] and at lunch time on the premier French national holiday, Bastille Day (14 July) in the year 2000, someone looking down on France from space would have actually seen that axis of the Hexagon. Thousands of French people – couples, families and groups of friends – congregated on the Paris Meridian, braving the rather poor weather that day and gathered to the borders of kilometer-long strips of traditional red and white checkered cloth. They were like tablecloths, laid down on the grass and on picnic tables. People set out bread, opened bottles of wine, unwrapped cheese and local specialities and ate lunch. This was a mass meal, called *L'Incroyable Pique-nique*, and a celebration at the start of France's new millennium.

---

[1] Seen from above, France has the outline of a hexagon, with two opposite corners to the north and the south. To the west, running roughly north-south from Brittany to the Spanish border, is the Atlantic coast of the Bay of Biscay. To the north west, from Brittany to the Netherlands, is the coast of the English Channel, or, as the French understandably prefer to call it, La Manche (the Sleeve). To the north-east, the hexagon's side runs along the borders of the neighbouring countries from the Netherlands, to northern Germany. To the east, the border continues southwards to the Mediterranean. The Mediterranean coast forms the south-east side of the hexagon, and the border with Spain completes it to the south-west. In French, the country of France is sometimes metaphorically referred to as the *Hexagone*, because of this shape.

Fig. 1 Map of France showing the Paris Meridian and places mentioned in the text

This "Incredible Picnic" was the human part of a more general, more permanent project in France, called *La Méridienne Verte* (the Green Meridian). In this project, French organizations, national, regional and local administrations, environmental groups, and estate owners planted trees along the same axis, across the Hexagon from north to south. It was the idea of the architect Paul Chemetov (1928-), whose major work in Paris is the courtyard of Les Halles. The plan was to plant 10,000 saplings, and the species were chosen appropriately to the region and the climate as it changed from the North Sea to the Mediterranean. In the north the trees were oaks, cedars and chestnuts, in the south olive trees. The idea was that in time the line of trees would become almost continuous, visible from space as the axis of the

Fig. 2 The location of the Méridienne Verte is marked by the side of the road on Route Nationale 152 near Manchecourt opposite to a stone obelisk erected to mark the position of the meridian of 1748 measured by César-François Cassini. Manchecourt is a village between Malesherbes and Pithiviers, south of Fontainebleau. There are few trees actually planted in this area to implement the Green Meridian concept! Photo by the author

Hexagon, making the Paris Meridian plain to see. From time to time as one travels on French highways across the Paris Meridian he or she will see by the side of the road a notice calling attention to the Méridienne Verte (Fig. 2).

WHAT WAS the reason for the choice of the location for the picnic and the trees? What is the significance of the axis of the Hexagon? What *is* the Paris Meridian?

On Earth every point is defined by two quantities – its latitude and its longitude. These are angles measured in degrees on Earth's near-spherical surface as seen from its center. If one knows the latitude and longitude to accurately to 1 arc minute (1/60 degree), then one knows where he or she is to about 1 nautical mile (about 2 km). Latitude is defined relative to the Earth's equator, in degrees north and south. Longitude is measured in degrees west or east of a north-south line on the Earth's surface that passes through both poles, but there is no natural line that is the zero point.

In order to help their ships navigate across the sea and to measure where they were in latitude and longitude, the maritime nations of the seventeenth to nineteenth centuries individually set up such lines through astronomical observation. Even landlocked countries set up meridians to help locate their cities and towns in latitude and longitude, although the need to define where the cities are in those terms is not so pressing. However, there is certainly a strong bureaucratic need to define areas of land, to identify legal ownership, and to assess taxes. An accurately measured north-south line can serve as the basis of an accurate map. Such a line is called a meridian[2]; the site of the "Incredible Picnic" was the meridian that runs through the astronomical observatory in Paris.

In the seventeenth century the need for accurate maps had become clear. In the sixteenth century, ships had been navigated by experienced mariners using "ruttiers" – written sailing instructions that described currents, winds, hidden hazards, landmarks, depths, anchorages, port facilities, and the nature of the sea-bed (Fernández-Armesto 2006). Charts were used as adjuncts but they were unreliable, in part because there were unexplored gaps and also because the representations of the geographic features of the shoreline were inaccurately placed. It was only as the accuracy of charts improved that navigators began to use them as the main navigational aid. Their accuracy increased as the result of improved technology, including more accurate measuring instruments, telescopic sights that made it easier to measure small angles, etc. In the seventeenth century, the Paris Meridian became the starting point to create accurate charts of France (not only the shoreline but also inland areas) and, for French sailors, to create accurate charts of the world. It was the origin of the French system of longitude, just as the meridian of Greenwich was the origin of the longitude system used in Great Britain. The Paris Meridian lies 2° 20′ of longitude east of the Greenwich Meridian.

The orientation of the French meridian north-south was based on the position of the stars. Most stars and the Sun rise up from the eastern horizon and pass across the sky, reaching their highest elevation before setting in the west (Fig. 3). Their position at their highest elevation, when they "culminate,"[3] marks south[4] – north is opposite this direction, east and west are at right angles. This is the basic way that astronomers determine the four cardinal points.

In practice and equivalently, south can be found by measuring the position of a star while it is rising and the position when it is at the same altitude while setting, taking the midpoint (Fig. 3). For example, south is the midpoint between the positions of a star or the Sun on the horizon when it rises and when it sets (with allowance for the motion of the Sun between these times). The direction of the star or Sun measured as an angle from north is called its "azimuth"; this is an Arab word that came into the

---

[2] In the English language the word *meridian* means the line running north-south through the poles, however determined and for whatever purpose. The French language distinguishes between *le méridien* and *la méridienne*, the former being the abstract line of points of the same longitude and the latter being a real line. *Une méridienne* can serve as a baseline for geographic measurements, as in this book, or it may be the visible line of a sundial showing noon local time. *Une méridienne* is oriented along *un méridien*.

[3] *Culminate* in astronomy or astrology means of a star "to reach its highest point in the sky" (from the astronomical meaning is derived the general meaning of the word "to reach an acme of development").

[4] The position of a star at its highest elevation lies in the south if you are in the northern hemisphere, as in France. In the southern hemisphere it would be in the north.

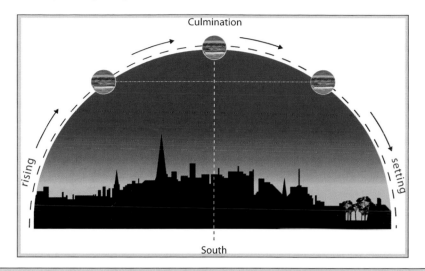

Fig. 3 Like all stars, Jupiter rises in the east and sets in the west. Seen from the northern hemisphere it "culminates", at its highest point in the sky, when it is due south. The direction of south is in practice found by measuring the position of a star at a given altitude as it first rises and then sets and taking the midpoint of these two directions

English language from the Islamic world, a culture that preserved and developed the classical knowledge of astronomy while Europe was in the Dark Ages.

In the other direction, the circumpolar stars, like the North Pole Star, define north; Polaris is not exactly at the North Celestial Pole, but circles around it and other circumpolar stars do the same. Astronomers measure the position of Polaris and other circumpolar stars at their western and eastern-most extensions from the Pole and determine the point midway between, which is the Celestial Pole, the direction of north. In the seventeenth and eighteenth centuries, the Paris astronomers mapped the geographic line to the north and the south of the Paris Observatory, progressively extending outwards through its grounds. This created the meridian and the French zero of longitude.

To relate places to the east and west of this meridian, the astronomers made simultaneous measurements of the positions of the same stars. As seen from a place to the east, a star comes to culmination earlier than when seen from a place to the west. The time difference represents the longitude difference between the two places (Sobel 1999), but how do you measure time difference? In modern times, the answer to this question is easy – you look at a clock. In earlier times before the invention of the chronometer, however, this question was more difficult to answer because seventeenth century artificial clocks were too inaccurate.

In contrast there are natural clocks, and in 1610 Galileo discovered that the planet Jupiter had satellites that revolve around it with accurate periods. The satellites are repeatedly eclipsed as they pass behind Jupiter and into its shadow. The satellites define a natural clock with the eclipses as the ticks. The timing of the eclipses was

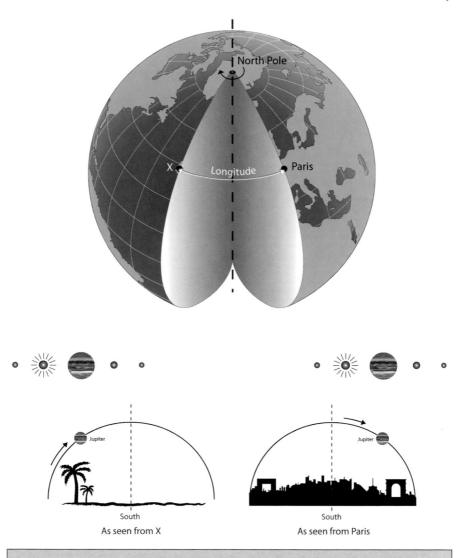

Fig. 4 An eclipse of one of Jupiter's satellites is viewed from Paris and from somewhere else on the Earth, X. At the moment when the eclipse happens, Jupiter is seen in different locations in the sky, and the eclipse occurs at different times of the day at the two places. The difference between the local times corresponds to the longitude difference between Paris and X

calculated and observed by the Italian astronomer Gian Domenico Cassini, who had become the director of the Paris Observatory. The natural clock of the motion of the satellites can be used to regulate a more convenient but less accurate artificial clock that can be used to observe the positions of stars and to determine longitude (Fig 4). These techniques were used not only to lay out the north-south meridian but also to determine by what longitude a place was offset from the meridian.

MAKING a map is a matter not only of determining the orientation of the map and the latitude and longitude of places on it but also of knowing the scale. This was done by surveying techniques, determining the distance between places on the meridian by physical measurements with standard measuring sticks and chains, and by triangulating with theodolites, or their historic equivalents. These techniques, called *geodesy*[5], were used to lay out the Paris Meridian and to relate it to places in the rest of the country and the rest of the world. The accuracy and the scope of the work were unprecedented. The result was to produce an accurate map of France for the use of sailors navigating along the coast and for administrative purposes, and these maps were the first maps of an entire country that were grounded in accurate scientific measurement.

In 1693 the need for the new maps was demonstrated by an outline map of the coastline of France after the first measurements had been made by the new techniques, compared in 1693 with the most accurate earlier map. Brest in Brittany had been moved to the east of its previously assumed position by more than a whole degree of longitude (about 140 km), and the older map had overestimated the area of France by a fifth (Fig 5).

The demonstration of the technique was timely and was extended to map not just France but its possessions overseas. In 1682 René Robert Cavelier (Sieur de La Salle) had just claimed the American territory of Louisiana for the French king (for whom the territory was named). He had the vaguest idea of the position and extent of this territory exemplified by his death in 1687. Setting out through the Gulf of Mexico for the mouth of the Mississippi River, he landed instead in what is now Texas, 1000 km away. His men lost confidence in their commander. They murdered him.

As we shall see, the techniques were extended even beyond the boundaries of individual countries. They were used to determine the size and the shape of the whole world, and even prove the scientific law that holds the whole Universe together. What an edifice to build on the abstraction of the Paris Meridian!

---

[5] You would think that *geometry* (from *geo-* earth and *–metria* measuring) would be the relevant word, and in early times it might have been, more so than now. *Geometry* was associated with the practical art of measuring and planning and was used especially in connection with architecture. But by the seventeenth century it had come to mean the properties of spaces, including lines and surfaces, and had taken on its present mathematical meaning. *Geodesy* (from *geo-* earth and *-daiein* divide) came into English to mean the measuring of land, and in the nineteenth century it developed its modern meaning of the branch of applied mathematics that relates to the figure of large areas of land and Earth as a whole.

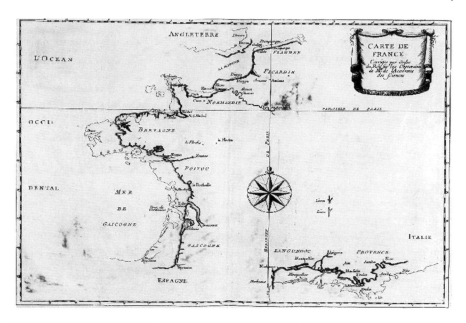

Fig. 5 Philippe de la Hire presented a map to the Paris Academy of Sciences in 1683 which showed the real outline of France compared with where it had previously been thought to lie. Brest lay 140 km nearer Paris than was thought until then. The area of France had been reduced by 20%. This "dramatic" map was published ten years later in 1693 as a 'Map of France Corrected by Order of the King.' This was the first map to show the Paris Meridian

Chapter 2
The Size of France

The groundwork to lay down the Paris Meridian was initiated during the reign of King Louis XIV (Debarth 1984). Louis XIV, the Sun King, came to the throne of France as a boy within a few days of his fifth birthday but took charge in his more mature years (18 years later in 1661). He established a brilliant court and was a patron of writers, artists, and intellectuals. In addition, he appointed Jean-Baptiste Colbert as a minister with wide authority to enhance the status of the King and to develop France thus restoring its fortunes after a weakening series of wars with Spain.

People Jean-Baptiste Colbert (1619–1683)

Colbert was born into a merchant family of Reims, and as a result of family connections he was appointed at a very young age to the war office as an inspector of troops, working his way up to become the personal secretary of the Secretary of War. At the age of 30 he became a minister of state and starting at 45 he became, in rapid succession, the superintendent of buildings, controller-general, minister of the marine, minister of commerce, minister of the colonies and minister of the palace, acquiring power in every department except war.

Colbert served as the minister of finance for 22 years under King Louis XIV. He improved the French economy, although Louis spent so much (on luxury and war) that France progressively became poorer. He ruthlessly rooted out corruption in government, introduced a fairer taxation system, and introduced numerous regulations to protect trade. He improved the French merchant marine as a means to facilitate international trade and built infrastructure to improve domestic trade – roads, ports and canals.

Colbert took a great interest in art, intellectual life, and literature. He founded the Academy of Sciences and the Paris Observatory, enriched the Louvre with hundreds of pictures, and supported many men of science and letters.

Colbert decided to make an inventory of France, a kind of French equivalent to the Norman survey of England that produced the Doomsday Book[6]. In 1663 Colbert sent out orders to the provincial governments to report on everything under their control, providing maps and other details of every imaginable resource like estates,

---

[6] The Doomsday Book (or Domesday Book) was a census of England completed in 1086 for William the Conqueror. William needed to administer the country he had just conquered in 1066 and to collect appropriate taxes. He sent out census takers to list all the villages and towns, their inhabitants, their land and their animals. The census still survives and is a valuable historic source. It became known as the Doomsday Book, because its data were supposed to last until the end of the world.

P. Murdin, *Full Meridian of Glory*,
DOI: 10.1007/978-0-387-75534-2_2, © Springer Science + Business Media, LLC 2009

water, crops, and churches. The documents were delivered to the royal cartographer, Nicolas Sanson (1600-1667), who compiled all the maps into summary works. The difficulties and inconsistencies in completing the survey confirmed what Colbert had suspected – that the maps were inaccurate.

Colbert decided to set up projects to rectify the situation and to tackle other scientific problems of importance to France. In the early seventeenth century, European scientists gathered in each other's houses for informal meetings. They networked and discussed the scientific works of each other and of the people who had sent letters to them describing what they had done. Even now scientists use terminology that reminds us of this time, for example calling some scientific papers "Letters to the Editor". In Britain these meetings led to the formation of the Royal Society of London in 1662. In France from 1665-1666 Colbert founded the Académie Royale des Sciences using a similar model, and he personally invited internationally known scientists to join the Academy in Paris.

One major difference between the Academy and the Royal Society was that Colbert offered the Academicians significant sums of money as salaries (1500 livres[7] per year) and for research (1200 livres), whereas the Royal Society's Fellows received nothing except the honor of being members. In the words of a recent President of the Royal Society, P. M. S. Blackett (1967): "King Louis XIV generously endowed his new Academy of Sciences in 1666 with funds to conduct experiments and make investigations. All that the Royal Society got from its founder, King Charles the Second, was a mace and his blessing." In return for financial support, however, the French Academicians had to work not only on projects that they themselves originated and carried out with their research funding but also on particular problems and projects chosen by the people who paid them. The British Fellows worked only individually and in ad hoc groups on what roused their curiosity. The Royal Society never executed national projects like the one described in this book.

Very soon after its foundation, a group of the Academy's new members made a proposal through Colbert to the King, in order to establish an observatory in Paris as a base for scientific investigations. The investigations were to be of a wide scope, not just what today would be called "astronomy." It was intended that the observatory would be the base for the whole Academy and its function a laboratory, not only for astronomy but also for meteorology and physics such as the determination of the speed of sound. Eventually its location was considered too remote from the center of Paris (it is a good two kilometers from Notre Dame!) The Academy was therefore provided with a new headquarters closer to the center of Paris and on the banks of the River Seine.

---

[7] The *livre,* literally a pound, was used as a unit of weight (roughly the same as the British and American pound). The name survives in French markets as meaning half a kilogram. The *livre* was also a unit of money based on the value of a pound of silver, but its specification varied across the country. In the seventeenth century, 1 *livre* could buy two pounds of butter or two chickens. In 1795 the Revolutionary Government replaced the *livre* with the *franc,* devalued by inflation at that time to 5 grams of silver, and now superseded by the single currency of the core nations of the European Union. the *euro.*

Places The Paris Observatory

The main buildings of the Paris Observatory (Fig. 6) were finished in 1672 by the architect Claude Perrault (1613-1688), although some work trickled on until 1683. Perrault was the Superintendent of Buildings under Colbert and is regarded more highly as an administrator than as an architect. The Observatory building was oriented accurately to the cardinal points and had a central rectangular court with octagonal towers to the east and west of its south side. Originally portable telescopes were taken on to the towers to make observations,

Fig. 6 Paris Observatory. The Cassini room, from which Cassini made the first observations of the Paris Meridian, is the high room under the clock above the main door. It runs through to the rear of the building, where sunlight was admitted through an aperture, to shine on the floor. The axis of the Cassini Room, and the axis of symmetry of the building, defines the original Paris Meridian. The statue is of J. J. Leverrier, the discoverer of Neptune. Photo by the author

but today these towers, suitably strengthened, carry permanent observing domes for the telescopes now installed there. A terrace surrounds the building and certain instruments were also taken on to the terrace for observations. The building was built without wood (to avoid fire) and without iron (to avoid disturbance to any magnetic measurements); some criticised it, however, for its unsuitability for astronomical observations because there is nowhere inside from which it is possible to view close to the zenith.

What has become known as the Cassini Room is situated centrally on the floor above the main northern entrance of the Observatory. It is a huge room, now reconstructed to contain pillars to support instruments and structures placed on the roof. Here the first leader of the Observatory, Gian Domenico Cassini, constructed the first Paris Meridian on a north-south line on the floor. In May 1682, the King visited this meridian to regulate his watch to accurate Paris time.

The meridian line in the Cassini room is now marked by a memorial erected by Gian Domenico's son, Jacques, to his father: "Having extended the meridian line that passes through the Paris Observatory in both directions, north and south, to the confines of the kingdom, it appeared necessary to the full perfection of the work to trace within the Observatory itself a meridian line that would be part of the one crossing the kingdom and, at the same time, would serve for the astronomical observations pursued vigorously there since its inception." The line is flanked by worn and rather crudely chiselled engravings of the zodiacal figures covered by transparent plastic overlays for their protection. The figures mark where the image of the Sun falls as it passes into the corresponding zodiacal sign.

The Observatory is decorated with other unique works of art. For example, on the ceiling of the Council Room, watched in perpetuity by portraits of the directors of the Observatory, is an allegorical picture (1886) of the Transit of Venus in front of the Sun, painted by Edmond Louis Dupain (1847-1933).

AS PART OF THE Observatory's program, even before its buildings were finished, the Academy took up Colbert's challenge to create better maps of France (Konvitz 1987). It charged one of its members, the Jesuit astronomer Abbé Jean Picard, with the measurements.

People Jean Picard (1620-1682)

Little is known of the early life of the Abbé Jean Picard although he started his working life as a gardener (or perhaps this is a misunderstanding of reports intending to say that he was a horticulturalist). Via training in the Church, he became a professor of astronomy in Paris and made several improvements in scientific instrumentation including measurements of the diameters of the Sun and the Moon, which since the angular diameters change with distance from the Earth, were used to help determine their orbits. He was considered in his time a scientist of the first rank, and in criticising the performance of the first leader of the Paris Observatory some historians have said that Picard ought to have been appointed instead.

Picard approached scientifically the tasks he had been given. The best available maps of France up to then were those by Sanson, yet as detailed, clear and elegant as these maps were, they were the product of office research and collation, not of field work. Colbert contrasted the method of their execution with the pioneering work by Dutch cartographers; they had based their methods on principles developed by Willebrord Snell.

People Willebrord Snell van Roijen (1580-1626)

Snell studied law in Leiden but became interested in mathematics and astronomy. He traveled throughout Europe and studied in Paris, and in 1613 he succeeded his father as professor of mathematics at the University of Leiden. He published *Eratosthenes Batavus*, in which he proposed the method of triangulation. He also worked on optics but did not publish in the field. His discovery of the sine law of refraction became known when a later prominent scientist, Christiaan Huygens, published it.

Snell had proposed the technique of triangulation as a way to measure the relative locations of points on the surface of the Earth (Smith 1986). The method, which is still the standard one, starts by the surveyors measuring a reference line between stakes on a flat plain using standard sized sticks or chains (Fig. 7). A third stake is placed somewhere else on the plain at a significant location. From each end of the line, a surveyor sights the stake at the other end as well as the third stake, measuring the angle between with a *theodolite*. The third stake is located relative to the other two by trigonometry of the triangle. From the reference points the surveyor can also sight natural vertical features in the landscape. Particularly crucial reference points can be made permanent by erecting pillars made of stone, sharpened like a pyramid to a point at the top to give the greatest potential accuracy to measurements of their position. All these features can be located relative to the others building a chain of triangles across the country – hence the term "triangulation".

Naturally, hill-top locations are important in the execution of the triangulation (Fig. 8). They are clearly visible from long distances and, on the contrary, from them large areas of the country can be seen below. The reference points that the early surveyors used on hills and mountains included not only chapels, windmills, thatched cottages, hunting lodges, ruined buildings and signals (iron basket-like

Fig. 7 The first stage in triangulation is to lay out a baseline on a plain. Surveyors use measuring sticks, laying them down alongside a straight line marked out by a rope or chain staked out on the ground. Here each measuring stick seems to be 4 metres long and is thus a two toise standard. All around the plain are suitable triangulation points – church steeples, towers, windmills, mountain tops, trees and the like – which will be sighted from the ends of the baseline. César-François Cassini de Thury (1744)

Fig. 8 From the plain, surveyors sight on hill-top locations, establishing a trig point from which they can sight back to further features in the landscape below. During the extended period of observations from a mountain location, the surveyors can make astronomical observations, timing astronomical phenomena to establish the latitude and longitude of the trig point, as well as its position relative to other places. Here, a surveyor consults a pendulum clock mounted in a religious shrine, while his colleague sights on a monastic building below. A third colleague on a hill-top across the valley sights on the same building to establish a link between the two hill locations. César-François Cassini de Thury (1744)

structures that could contain signal fires), but also (in uninhabited areas) natural features like large rocks and solitary trees. For key reference points to be used repeatedly over time, which must be unambiguous and clear, surveyors erect artificial pillars as triangulation points (or "trig points" for short). These pillars remain characteristic features of a hill-walk in a mountain landscape in France or Britain, although they are all too often subjected to vandalism and in stressful times have been used as target practice by invading forces of artillery.

In the valleys, surveyors sought tall, thin markers like the spires of churches, weather vanes, and the corner turrets of towers. Church spires in particular are likely to remain a permanent fixture. Moreover, they usually are at the centers of communities where the maps were likely to be of the most practical use. In addition, they are usually a convenient tapering shape – if the church had a square-cut tower, then a triangular tent-like shape could be placed on top of it at its mid-point. To provide an intermediate baseline that helped establish a more important, more permanent reference point, the surveyors used temporary structures such as a hole in the roof of someone's house or a sentry box. (All the examples quoted in these paragraphs are found as trigonometric points for the Paris Meridian surveys).

To survey the longer sides of the triangles, say over 100 km from one mountain peak to another, the geodesists observed at night, using a lantern with an optical reflector to concentrate the light into the right direction along the arm of the triangle. This had an advantage because at night the atmosphere over such long distances was settled more than during the heat of the day. The images were clearer, refraction of the rays in air of different densities was less confusing, and the angles could be measured more accurately.

Snell's techniques were demonstrated by relatively small-scale projects by him and other Dutch cartographers. The first survey by triangulation is regarded to be Snell's survey prior to 1615 of the Netherlands from Alkmaar via Amsterdam, Leiden, Utrecht, Dordrecht and Breda to Bergen-op-Zoom, accumulating an overall distance of 120 km. This revealed the considerable demands, not just for the survey and the measurements, but also for subsequent calculations. Not until the invention of the decimal point in arithmetic notation and of logarithms to replace long multiplication and division by addition and subtraction[8], as well as the invention of accurate instruments to measure angles, was it practical to make a large-scale cartographic survey of an entire country.

The Academy and Picard took up Colbert's challenge with enthusiasm, using instruments which Picard had made and which were the most accurate of the time. Picard set out to survey a line, on the same scale as Snell's, from Malvoisine, a country building southeast of Montlhéry 30 km south of Paris, to Sourdon, 20 km south of Amiens, in total a distance of some 150 km along the north-south line through Paris.

PICARD'S FIRST TASK as he began in 1668 was to lay and measure a fundamental baseline on an arc of the meridian. The baseline ran from Villejuif, just south of Paris, to Juvisy, 11 km further south, across the flat plain near Fontainebleau (right across what is now Paris's Orly airport). Picard determined its orientation astronomically by observing star positions and laid it accurately north-south.

To measure the length of the fundamental baseline, Picard used wooden rods at the standard length of 1 toise[9]. Two assistants were each given a one-toise measuring rod and five pegs. Starting at the first surveyor's stake, the first rod was laid on

---

[8] The first person to use a decimal point to separate the whole number part from the decimal part was the Edinburgh mathematician, John Napier, Laird of Merchiston in Scotland, in a book called *Mirifici logarithmorum canonis description* (1614), where he also explained his invention of logarithms (other contenders for the title of inventor of logarithms include the Swiss mathematician Jobst Burgi, who was credited as such by no less an authority than Johannes Kepler): "Seeing there is nothing (right well beloved Students of the Mathematics) that is so troublesome to mathematical practice, nor that doth more molest and hinder calculators, than the multiplications, divisions, square and cubical extractions of great numbers, which besides the tedious expense of time are for the most part subject to many slippery errors, I began therefore to consider in my mind by what certain and ready art I might remove those hindrances. And having thought upon many things to this purpose, I found at length some excellent brief rules to be treated of (perhaps) hereafter. But amongst all, none more profitable than this which together with the hard and tedious multiplications, divisions, and extractions of roots, doth also cast away from the work itself even the very numbers themselves that are to be multiplied, divided and resolved into roots, and putteth other numbers in their place which perform as much as they can do, only by addition and subtraction, division by two or division by three." The techniques produced workable tables of logarithms from 1624.

[9] A *toise* was the French unit of length before the meter, and was about 5 centimeters short of two meters long.

Fig. 9 The southern end of Picard's and Cassini's baseline at Juvisy-sur-Orges, near Orly Airport, is marked and commemorated by a stone obelisk, called La Pyramide, which stands in a jumble of suburban buildings and street furniture. (Photo by the author)

the ground, its side marked by a peg. The second assistant laid his rod at the end of the first rod, slipping a clip from the second one to the first to hold them in place and marked his own rod with a peg. The rods were thus leap-frogged until the pegs were exhausted and this marked out part of a baseline of ten toises length. This pegging system made it almost impossible to lose count, and the section of ten toises was noted down on paper. This process was repeated 500 times until the fundamental baseline was staked out.

Places The obelisks at Juvisy and Villejuif

The southern end of the baseline at Juvisy-sur-Orges is commemorated by a stone obelisk (Fig. 9), called La Pyramide, next to Route Nationale 7 and lying on the east side of the road at the intersection of the D25 to Athis. It was built in 1740, and the tapered column is set on a square plinth that stands on a small mound. The plinth carries a modern geodesic marker of iron; the whole monument is made of layers of a soft, yellow stone that have weathered to gray and reaches some 8 meters high at an ornamental stud. It carries this inscription:

PYRAMIDE DE JUVISY
EXTREMITE SUD DE LA BASE GEODESIQUE
DE VILLEJUIF A JUVISY
1670 (PICARD)
1740 (J. CASSINI ET LACAILLE)

–

PROPRIETE DE L'ACADEMIE DES SCIENCES

(Pyramid of Juvisy, southern end of the geodesic base from Villejuif to Juvisy. 1670 (Picard), 1740 (J. Cassini and Lacaille). Property of the Academy of Sciences.)

The Pyramid is surrounded by a suburban Paris of North African culture with the Boulangerie (Bakery) de la Pyramide, Garage de la Pyramide, etc, adjacent to it and cheap stores and run-down apartment buildings crowding onto the surrounding streets. This suburb was a location in Virginie Dispentes' notorious film *Baise-Moi* (2000) in which a drug dealer is murdered during inter-gang warfare. Today it is not easy to recognize in this blighted area the scene painted by J. Cassini in his description of the triangulation point he used nearby: "Juvisy: It is the first tree of the avenue that is to the left in going from Paris to Fontainebleau, after the gate of the park of Juvisy."

Further south along RN 7 is a curious castle-like building that was the observatory of Camille Flammarion (1842-1925), a well-known, popular writer of astronomy and founder of the Astronomical Society of France.

The northern end of the baseline is commemorated by a similar obelisk at Villejuif, smaller and more slender than Juvisy, surmounted by a vertical iron rod and without an inscription (Fig. 10). It stands in the garden of 157 bis Avenue de Paris, which can be reached from RN 7 just north of the intersection with D61. The obelisk stands on a knoll high above the level of the road from where it can be readily seen. It can be approached more closely by a flight of stone stairs marked "Privé" and can be viewed over the garden fence. There is a marker for the Méridienne Verte at the obelisk's foot, and some saplings were planted nearby in 2000 as part of the Millennium celebrations.

To lay the rods in a straight line along the chosen route from Villejuif to Juvisy Picard must have established a new, or straightened a pre-existing track through the area's woodlands. This route must have been useful for communication from the south into Paris, and the track was evidently maintained and developed by Frenchmen who came through the area after Picard left. His track became the Avenue de Paris in Villejuif and, eventually, Route Nationale 7 through Orly to Juvisy. Picard would have been baffled and astounded if he had known his track would one day become a four or six lane highway tunneling under the airplane taxiways of an airport.

Having laid out his measured baseline, Picard triangulated from its ends to landmarks in the surrounding landscape. From those landmarks he triangulated to others ultimately heading right across Paris to the north and a little towards Bourges to the south. Of course not all potential landmarks were equally useful – to measure a triangulation triangle it must be possible from any one trig point to see each of the other two. To establish this it is necessary to check out the potential landmarks and check

Fig. 10 The northern end of Picard's and Cassini's baseline is marked and commemorated by an obelisk at Villejuif, standing in a suburban garden, with trees newly planted nearby to mark the Méridienne Verte. (Photo by the author)

intervisibility (it would be disappointing to climb to the top of a mountain with loads of equipment and find that there was a peak in the way of the line of sight to one of the other landmarks). Typically, the landmarks alternate from one side of the meridian to the other so that when candidate landmarks are identified their position was roughly estimated. In this way Picard drew up his network of triangulation stations.

After choosing the trig points, Picard or his colleagues made many measurements from each station and documented the features to be transcribed on to the maps. They often chose churches as triangulation stations. As well as being highly visible, permanent and suitably shaped, churches had particularly practical advantages. They were likely to be near lodging and other infrastructural support that the scientists could use during the measurements. They were high, so that the terrain around could be reconnoitered easily, and each was in the charge of an educated person, the curé, who could understand requests, provide advice and effect introductions. The first task of a surveyor on arriving at a new place would be to seek out the curé or some other local lord and get this person to name the features of the landscape that could be seen from church tower – the villages, the towers, the woods, the mountains, the windmills, the abbeys, … This might well have been the first time that these names had been recorded, and the surveyor was at the mercy of his informant. This person might have some hidden agenda; for example an egotistical landowner might want to name features after himself, or give names that tended to establish his possession of something that was held in common or whose ownership was disputed, or not give names to features that he wanted to lose (like a ruined farmhouse, to which occupancy rights might later be claimed).

At a minimum, the surveyors had to carry to the station their surveying instruments to measure the angles of the triangle. These surveying instruments had to be set up and levelled by means of screw adjustments on their feet against a reading from a plumb bob. When correctly positioned, this plum bob hung over a fiducial mark when the rotation axis of the instrument was vertical. To be successful, the surveyors had to wait for clear weather and make several observations at different times of the day and at night, especially for distant stations. Night-time observations were made by sighting onto a lantern at the distant station; someone had to go there to erect the lantern and its optical reflector and to light the lantern at a pre-arranged time.

The surveyors also measured the latitude and longitude of some selected stations using astronomical measurements. Such observations, though, depended on good weather and there being suitable astronomical phenomena to measure (e.g. eclipses of Jupiter's satellites). The eclipses had to be timed not only at the station but also at the Paris Observatory so that the local time interval between them could be determined in order to measure the difference in longitude, and the observing programs at the two stations had to be coordinated. Time is measured by a clock so the surveyors had to carefully carry a clock to the site where it would then have to be set up, leveled, and regulated. Pendulum clocks had just been invented (see below) and were an enormous advance in accuracy over anything used before, yet they were over a meter high and extremely delicate. Also, the clock had to be regulated by repeatedly and over some days observing the transit of the stars in their 24 hour sidereal motion.

All this preparation takes time – days, weeks, even months. If the station was in a village, Picard could negotiate to buy accommodation at an inn, or he might be fortunate enough to be given hospitality by a priest of a church or the owner of a château whose spire or tower had been chosen as a landmark. Picard would need a place to set up and keep his instruments. If nothing suitable was immediately available then he would have to build a wooden shed, robust and spacious enough to provide shelter from the wind for the observers, especially during the cold night-time observations (Fig. 11). Somewhere there would have to be some storage for the instruments, a place

Fig. 11 Two of Cassini III's observers make stellar observations with an octant mounted in an observing shelter. One lies in a reclining chair to sight on a star while the other reads an angular scale with a microscope. It seems a little unlikely that they could have managed with such a small hole in the roof through which to view the sky. César-François Cassini de Thury (1744)

to cook, and a bed for sleep as necessary. During all this Picard was supported by servants who traveled with him, and about whom we know little or nothing.

One trigonometrical station in particular played a continuous role in the history of the Paris Observatory and the surveys. It was erected next to a windmill (the Moulin de la Galette) on the hill of Montmartre, directly north of the center line of the Observatory building. The pillar originally served to calibrate the instruments of the Observatory where it was visible from (trees and houses now interfere with the line of sight). In French, this is called a *mire*[10] and appropriately the Rue de la Mire is nearby. Picard placed the first *mire* here, a wooden pole to serve this function. Its position to the north of the center of the Observatory building was established by a survey in 1673, and a substantial wooden post was put at the same place in 1675. However in 1683 Cassini and La Hire determined that its position was two feet west of the meridian, an error of azimuth of about 0.5 arc minutes, or 1/120 of a degree. This post was further superseded by a stone *mire*.

To measure the angles between landmarks, Picard used a seventeenth century equivalent of a theodolite, a large quarter circle (quadrant) divided into angular measure that could be read to about 1 arc minute. It was 1 meter in size and mounted horizontally, and a small telescope with cross hairs (Picard pioneered the use of spider webs) was pivoted at the center in order to sight landscape features.

---

[10] *Une mire* is a target for calibration (the same word is used in modern times to mean a "test card" for TV transmissions).

The telescope was moved from one feature to the other, and as viewed from the telescope, the angle between two features was the angle entered into the trigonometric calculations.

Places The *mire* at Montmartre

Picard put the original trigonometrical station on the hill of Montmartre in 1673. The present stone obelisk dates from 1736 and was placed here by Jacques Cassini. Originally it was inscribed:
L'an M DCC XXX VI cet obélique a été élevé par ordre du Roi pour servir d'alignement à la méridienne de Paris du côté du nord. Son axe est à 2931 toises 2 pieds de la face mériodionale de l'Observatoire.
"In the year 1736 this pyramid was erected by order of the King to serve to align the meridian of Paris on the northern side. Its axis is 2931 toises 2 feet from the southern face of the Observatory."
As a result of weathering, only the date can now be read. Presumably in the eighteenth century the houses and trees less obscured the view of the obelisk from the Observatory. Now, the obelisk stands overshadowed by a clutter of urban outbuildings placed without any sympathy for its importance and reducing it from its former glory.

The survey starts off with a measured baseline that can then be extended by measuring the angles of triangles to the most distant point. As a check on the process and to determine the accumulated error, it is best if the survey includes at check points the extremities of the triangulation line or another baseline, whose length is determined not only from the trigonometry but also from direct measurements. According to the check, Picard's survey was accurate to 5 meters. Nearly 100 years later, though, it was concluded that this was a misleading impression of the accuracy and there must have been canceling errors.

IN 1669 while Picard was working on his first survey in the field, Colbert, in the name of the King, invited the Italian astronomer Gian Domenico Cassini to visit Paris and work at the Observatory. This was the beginning of a long collaboration between the two men.

People Gian Domenico Cassini (1625-1712)

Gian Domenico was known as Cassini I (Fig. 12) to distinguish him from his son, grandson, and great-grandson (all served as senior figures in the Paris Observatory). He was born in Perinaldo, near Naples. Attracted to astrology in his youth, he became Professor at Bologna studying the rotation periods of the planets and the satellites of Jupiter. In the 1650's he reworked and corrected the construction of the accurate sundial, or *meridiana*, in the Church of San Petronio in Bologna to measure the motion of the Sun, timing and dating the position of the Sun's transit and image across an accurate north-south line (see Chapter 4).
When he moved to France in 1669 and was appointed to the Paris Observatory, Cassini had intended it as a limited visit, but his stay became permanent, his intentions changed by a woman. He married this woman, Geneviève de Laistre, daughter of the lieutenant general of the Comte de Clermont north of Paris. Her dowry included the Château de Thury, where the Cassini family lived for years thereafter. He took up French citizenship and altered his name to Jean Dominique, becoming fully adopted into the French establishment.

When Cassini arrived in Paris he erected and used a telescope of 17 feet (5.2 m) focal length given to him by the instrument maker Giuseppe Campani and with this made more accurate measurements of the Sun's rotation. The excellence of the

Fig. 12 Jean Dominique Cassini painted by Durangel in 1879, after an older original. Cassini looks across a celestial globe out to the Paris Observatory, with one of his long telescopes mounted on the roof. ©Observatoire de Paris

telescope impressed the court and the astronomers of the French Academy, particularly Colbert. Ultimately a 34-feet (10.4 m) focal length telescope was ordered from Campani and installed in the Observatory.

On the second floor of the completed Observatory building, Cassini created a room for observations of the Sun's positions. Today this is known as the Cassini Room and it is here Cassini repeated his experiments in Bologna and constructed another *meridiana*, or sundial (see Chapter 4). Cassini timed and measured the position of a shaft of sunlight that had passed through an aperture high on the south side of the room and fallen on a north-south line on the floor. This line was the first Paris Meridian, right on the building's line of symmetry.

Jean Dominique Cassini was not actually named Director of the Observatory, nor was his son Jacques Cassini (1677-1756 and known as Cassini II) who succeeded him. The Observatory was under the control of the Academy and its members could use its facilities to carry out any work they pleased if they could gather funding for it. In this capacity, they would not have regarded themselves as being directed by anyone. The visiting scientists brought their own equipment to the Observatory for their experiments and took it away when the experiments were finished. In contrast to an observatory today where its equipment is bought, built, and maintained as a common facility for all visiting scientists, the Observatory held minimum equipment for common use. This situation was similar to that of John Flamsteed, the first Astronomer Royal at the Royal Observatory in Greenwich, and during his tenure there from 1675 to 1719, he bought his telescopes and clocks out of his own salary. When he died, the state claimed ownership of his instruments as crown property. In response, his widow sent a servant to steal the telescopes and clocks away during the night, and sold them to provide financial support in her widowhood. One long-case clock for astronomical use had a pendulum 2 meters long but this was altered to the conventional size for a domestic clock to make it more saleable. Only recently was this clock restored to its place in the Octagon Room at Greenwich.

It was not until 1771 under the rule of Louis XV that the post of Director materialized, notably through César-François Cassini de Thury (1714-1784), known as Cassini III, the second son of Cassini II, and grandson of Cassini I. The final astronomical member of the Cassini family was Jean-Dominique Cassini (1748–1845), son of Cassini III, named in honor of his grandfather and known as Cassini IV, also formally a Director of the Paris Observatory and ennobled as the Comte de Thury. Formalities aside, the Observatory was essentially under the successive authority of Cassinis I, II, III and IV and they were, whether explicitly called that or not, the Observatory's first four Directors. Collectively these four men are known as the Cassini dynasty.

Places Château de Thury

The town of Thury-sous-Clermont (Oise) is directly associated with the Cassini dynasty. Cassini I took the title of Lord of Thury and Fillerval and the Château de Fillerval was the country residence of Cassini II to Cassini IV. Cassini III, who was born in the old castle, took the name Cassini de Thury after the town. He demolished the old castle of his birth but died before he could complete the present one, largely built by Cassini IV. Cassini IV was officially ennobled as the Comte de Thury and retired here from his life of science, lucky enough to keep his head after he was deposed from the Paris Observatory's directorship in 1793 by the revolutionary government. At the Château de Thury, he wrote memoirs, poetry, and was active in local politics. He died here in 1845, and his tomb is in the church's cemetery. The castle was damaged by fire and also by the Second World War but has been restored and is presently a business school.

In his time in Paris, Cassini I mapped the Moon and made observations of Jupiter's satellites to use as natural clocks. He observed the eclipses of the satellites and progressively drew up more accurate tables for predicting them. He also discovered that Saturn's rings were separated in two by what is now known as Cassini's Division. The European Space Agency's spacecraft, CASSINI, which arrived at Saturn in 2004 to explore the planetary system, is called such after him. He made other discoveries of physical astronomy (i.e. the nature of celestial objects), but this work of brilliant individual studies was regarded with mixed feelings by some Academy astronomers, who thought that the Observatory should concentrate on more fundamental, large scale works, or "big science" projects like the survey to establish the Paris Meridian.

As soon as Cassini arrived in Paris, he was enlisted to work with Picard to extend the meridian from its base on the plain near Fontainebleau. Between 1669 and 1670, Picard and Cassini measured the long baseline running north-south, south of Paris from Juvisy to Villejuif, and completed the survey northwards to Amiens. They measured the length of a degree of latitude across Paris at 57,060 toises, corresponding to 111,092 meters (the modern value for the average length of a degree of latitude is 111,220 meters). In 1671, their work was published as *Mesure de la Terre* in 1671 and resulted in a *Map of the Paris Region* published in 1678.

IN 1671-1672 PICARD took on the practical task of surveying, intending to shed light on the most important scientific theory under development at the time – some say the most significant scientific theory ever – the theory of gravity. He visited Brahe's ruined observatory, Uraniborg ("Castle of Urania"[11] or "City of the Heavens"), at Hven, in order to measure its position. The reason the position of Uraniborg was so important to Picard was that, if its accuracy was known, observations that had been taken there a century before could be related to other measurements, particularly those being taken by Picard and his colleagues at the Paris Observatory.

Uraniborg was the observatory of the Danish astronomer Tycho Brahe and its conception was triggered by the appearance of a new star. In 1572 Brahe was riding in his coach going home from a convivial dinner in Copenhagen when he noticed some peasants marveling at the sky. It was a bright star that had appeared in the constellation of Cassiopeia, where no star had been seen before. This was remarkable because it seemed to be a change in what was believed to be the "unchangeable" celestial heavens. Brahe was stunned that this idea, which was a key concept for Aristotlean philosophy. The model of the Universe proposed by Ptolemy that the Moon, the planets and the Sun revolved around the Earth within the sphere of fixed stars was be wrong. According to this philosophy, the planets moved and changed brightness because they were near to the Earth where everything was in flux, altering progressively until death. By contrast, the stars were at great distances next to heaven and were perfect and unchanging. What the peasants had seen seemed to be a star, and it had changed from being invisible to being one of the brighter stars in the sky.

To see if the star was really in the celestial sphere, Brahe made careful measurements over a year of the star's position relative to other stars. He used a version of cartographers' surveying techniques that triangulates on heavenly bodies and determines their distance. Brahe was at his observatory on the Earth in Copenhagen but was being carried around the Earth as it rotated in its daily motion. If he sighted on the star at 6 p.m. as the Sun set and then again at 6 a.m. as it rose, he would have been carried round the Earth by one half of a rotation. The sightings effectively established the angle to the star as seen from these two positions, called the *method of diurnal parallax*. Tycho established that the star never moved and as far as he could measure its parallax was zero. This proved that the star was more remote than, say, the Moon. In the book *De Stella Nova* ("Concerning the New Star")

---

[11] Urania was the classical muse of astronomy.

Tycho published his observations and his interpretation that the new star was truly a star and that the heavens had changed. Modern astronomers would agree with this; in their terminology the star was an exploding supernova.

People Tycho Brahe (1546-1601)

The astronomer/astrologer and alchemist Tycho Brahe, the survivor of a pair of twins, was born into a noble family in Denmark. He was educated by his uncle and later at the University of Copenhagen. An eclipse in 1560, and the fact that it had been predicted, inspired him to turn to astronomy. At the age of 17 he observed: "What's needed is a long term project with the aim of mapping the heavens conducted from a single location over a period of several years." These words consequently mapped out his future scientific work. After a student-related squabble, Tycho lost part of his nose in a duel and for the rest of his life he wore a replacement nose allegedly of gold (post mortem examination of his remains in 1901 suggests that it was really of copper). He traveled throughout Europe until in 1571 he set up an observatory in Copenhagen. In 1572, he observed the supernova of 1572 and was financed by the King to establish an Observatory at Uraniborg. He moved to Prague in 1599 where his patron was Rudolf II, the Holy Roman Emperor. He developed a model of the solar system, which had the Sun orbiting the Earth while the other planets (except the Earth) orbited the Sun. It was a half-way house between the Ptolemaic Theory that all the planets and the Sun orbit the Earth and the Copernican Theory that all the planets orbit the Sun, and only the Moon orbits the Earth. He died in 1601, after straining his bladder by retaining urine for the duration of a banquet out of a mistaken sense of propriety. However, some claim he died from mercury poisoning, whether self-administered as medicine or accidentally ingested during astrological experiments. Some say it was administered with the intention to murder but that is never going to be certain. His records of the motion of the planet Mars were passed on to his pupil Kepler, who used them to discover the laws of planetary motion.

The book attracted the attention of King Frederick II of Denmark and Norway. He financed Tycho to build next to Uraniborg, an observatory containing underground chambers with more stable foundations and temperatures for the instruments, which Brahe called Stjerneborg ("Star Castle"). Tycho occupied the complex until 1597 when Frederick's successor, Christian IV, came to the throne. Tycho then moved to Prague where he had the support of Emperor Rudolf II. The observatory was destroyed after he left and all that remains today are the ground works, fitted with a modern roof as a vistors' center.

At the Uraniborg observatory, Tycho had installed astronomical instruments he had made himself – not telescopes because they had yet to be invented, but sighting instruments that measured the positions of stars and planets. He created these instruments to take what were then the most accurate measurements of the stars and the positions of the planets, ultimately determining their positions to a fraction of an arc minute. His pupil Johannes Kepler (1571-1630) used Tycho's measurements to establish the planetary laws of motion. The measurements were also useful in further investigation of the accuracy of these laws and determining any departures from them, if Brahe's measurements could be linked to observations being carried out later, in Paris.

Picard gathered the data to do this, using the eclipses of Jupiter as clocks that he observed from Hven and also at Paris. Each eclipse progressed over many minutes, but there were two principal opportunities to make a decisive observation of the timing to an accuracy of seconds. The satellite gradually approaches the column of shadow behind Jupiter, its approaching side touches the edge of the shadow (first contact) and the satellite starts to fade. When its receding side enters into the shadow the satellite disappears (second contact). This is the most decisive moment to time, the disappearance of the *last speck*. As the satellite exits the shadow the forward side of the satellite emerges first (third contact). Although this moment (*first speck*) is as well defined as the second contact, it may occur at an unexpected time, given the uncertainty of the predictions, and catch the observer by surprise; hence this contact is the less accurate to time than the second contact. The moment when the satellite has fully emerged from shadow is the fourth contact. The first and the fourth contacts are not so decisive as the second and third, because they show to the eye as the beginning and end of a slow progression of brightness, rather than as abrupt appearance and disappearance.

Ideas Kepler's Laws

First Law: The planets move in elliptical orbits.
Second Law: They move such that the line drawn from the Sun to a planet sweeps out equal areas of the orbit in equal times.
Third Law: The distance and period of each planet are related to all the others in what is known as the harmonic law, that the square of the period of a planet is proportional to the cube of its distance from the Sun.

It is possible by eye to time an eclipse of the closest, fastest moving satellite, Io, as it passes into Jupiter's shadow to a few seconds, as measured on a clock at each observatory (van Helden 1996). The same eclipse observed simultaneously in two locations at different apparent times yielded the observers' longitude difference to an accuracy of better than 1 arc minute, the equivalent of a couple of kilometers.

During his stay in Hven, Picard worked with the Danish astronomer Ole Rømer (1644-1710) returning to Paris with him. In Paris, Rømer observed the satellites of Jupiter and noticed that the eclipses of its satellites occurred earlier than scheduled when Jupiter was close to the Earth and behind schedule when Jupiter was furthest away. Following this observation in 1676 he deduced that light travels at a finite speed and measured it very accurately and this discovery helped improve the predictions of the eclipses. This was important because it was not usually possible to compare observations of an eclipse from one place with observations of the same eclipse from another; it was more feasible to compare observations with a prediction. When the observed eclipse was timed relative to the tables of predictions, the longitude difference could be in error by several minutes of time, perhaps a degree of longitude, due to the inaccuracy of the predictions. Curiously Cassini did not take up the advantage of extra accuracy that Rømer's work had given him, and it was left to astronomers at the rival observatory at Greenwich to fully to exploit Rømer's discovery.

Observing Jupiter's satellites as a means to determine one's position was feasible from land but was not a practical way to determine time in order to navigate a ship. In order to view the eclipses of the satellites, a telescope was necessary. Telescopes of the time were long and thin (since opticians could make only small lenses), and the lenses could only produce good images if they were weak with shallow curvature; such weak lenses throw an image at a long distance. On the heaving deck of a ship the telescopes wavered in the hands of the observer, therefore the observing navigator could not easily hold such a telescope steady enough to view the planet continuously for the many uncertain minutes of the predicted eclipse, see it and time it accurately. To correct this, some effort was put into making a pendulum arrangement of a bosun's chair for the observer to sit in, suspended with a system of ropes and pullets from a pole or mast. The idea was that the observer would stay still while the boat rocked below him but this was not the experience. Eclipses often happened while the telescope was off target, and when the eclipse was seen the observer was too unsteady to decisively see decisively the moments of contact between the satellite and the shadow so as to time the eclipse accurately.

Even ashore on steady dry land, the telescope was liable to tremble in the slightest wind. It was, however, possible during calm weather to observe the eclipses from shore with the telescope supported by strings on static poles. This gave the longitude of the telescope's observing station, and, with the latitude observed from star transits across the meridian in the sky, the means to map a coast line.

In 1690 as an aid for sailors and explorers, the Academy began publishing the *Connaissance de Temps*, literally the 'Knowledge of Time' and the equivalent of a *Nautical Almanac*. Its annual issues contained predictions of the eclipses of the satellites of Jupiter, for the purposes of determining the time as measured at Paris, and instructions on how to do this:

For finding the longitude of any place on Earth it suffices to observe any immersion or emersion. One compares the true time of the observation with the hour and the minute of the same immersion or emersion calculated for Paris, or observed there the same day. The difference in time, reduced to degrees minutes and seconds will be the difference between the meridian of this place and the meridian of Paris.

USING TECHNIQUES that he had developed in Hven, Picard and Cassini mapped the shore-line of France, determining the latitude and longitude of capes, ports and other cities. Picard chose Philippe de la Hire to help them both.

People Philippe de la Hire (1640-1718)

Philippe de la Hire (Fig. 13) was the son of a painter and sculptor and this background gave him geometry training. While attempting a difficult sculpture, he developed a new method for constructing conic sections. His mathematical interest soon led to a curiosity in astronomy and physics, and in 1682 he became a member of the Observatory in Paris where he took part in observing programs of a wide variety. He was very productive and wrote and lectured extensively but produced no lasting innovations in science

In 1672-4 Picard went to Touraine, Languedoc and Lyon and in 1672 Cassini traveled to Provence. In 1679 Picard and La Hire traveled to Nantes and Brest to map western France and in 1680 to Bordeaux. In 1681-2 they worked eastwards along the northern coast, Picard in Normandy and La Hire near Calais. During this time, the Paris Meridian had been accurately measured across France from the northern to the southern coasts, and the length of the axis of the Hexagon had been determined but the task still remained to draw it on the map. In 1681, Picard proposed a plan to the Academy to extend his triangulation survey to cross the whole of France. He noted that parts of France that had been accurately surveyed near the

Fig. 13 Philippe de la Hire (self-portrait). ©Observatoire de Paris

coast and across the Paris basin were separated by vast areas that had not. Therefore, although the position of the coast line was now known, individual maps of the interior could not be assembled within the coast into a whole.

The extent of the problem was demonstrated by the map presented (possibly by La Hire) to the Academy in 1683 which sketched in the outline of France as independently established from surveys by Picard, Cassini and La Hire. The map (Fig. 5) compared the new outline of the coast with the earlier map of the coast by Sanson. It showed that Brest lay 6° 54' west of Paris, not 8° 10' as originally thought. It had been moved about 140 kilometers east, and located accurately to the kilometer. The survey reduced the size of France by 20%, from 31,657 square leagues as shown on Sanson's map to 25,386 square leagues by the new measurements[12].

The King is said to have wryly remarked that he had been ill-served by the astronomers whom he had supported. They they had made his kingdom smaller; in fact he had lost more of France to the astronomers than to his enemies. This "dramatic" map was published ten years later in 1693 as a *Map of France Corrected by Order*

---

[12] 1 league was 4444 meters and 1 square league was 20 square kilometers.

of the King with a superposition of the line of the accurate survey of the French coast on Sanson's map showing the startling differences. This was the first map to show the Paris Meridian.

Having mapped France, Cassini then mapped the world using the same techniques. He trained French explorers and Jesuit missionary priests in the techniques of determining latitude and longitude. When they reached their assigned destinations, they measured where they were and sent the results back to Paris. Cassini used the measurements to position places on a huge world map laid out on the third floor of the Observatoire de Paris in a circle about 10 meters in diameter, drawn on the floor with the North Pole at the center and with a concentric and radiating system of lines of latitude and longitude at ten-degree intervals. Using the measurements of latitude and longitude from around the world, Cassini added far-flung cities to the map like Québec, Santiago, Lisbon, Venice, Cairo, Siam, India, Canton, and Peking. "When he visited the work in progress, Louis XIV could stamp on the world and pinpoint places with toecap accuracy." (Fernández-Armesto 2006) In 1696 Cassini published the result of all this work and it became the first accurate world map.

It had taken 20 years for the astronomers and geodesists at the Paris Observatory to confirm Colbert's original suspicions about the maps of France and to show the scale of the problem. It was indeed substantial! But while the work had not resulted in an accurate map of the whole of France, it had provided evidence of the extent of the solution to the problem– the equipment, the numbers of people that were needed, the training that would provide their skills, and the time-consuming expeditions.

BASED ON THE experience of the trial surveys, in 1681 Picard defined a plan to triangulate along the whole of the meridian across France towards the north and the south and measure arcs in lines of latitude to the east and the west. This framework (or *chassis général*) would serve as the secure basis of smaller maps of the interior, linking them securely together. However Picard died in 1682, and in 1683 his plan was taken up by Cassini I with Colbert's backing. In the summer of 1683, the plan was approved by the Academy and assigned resources by Colbert. Cassini I led the southern expedition to Perpignan in the south and La Hire went to the north.

In 1683 the plan suffered a grievous set back. Colbert died, and his successor in charge of the Academy, François Le Tellier, Marquis de Louvois, was not so enthusiastic. By the end of 1683 the meridian had been surveyed from Mont Cassel in the north, near Calais, to Montluçon near the Massif Central, but field work was suspended for the winter because the snows in the Auvergne and in Limousin rendered several key stations inaccessible. Louvois recalled the expeditions to wait for more favorable weather in the spring. He never reactivated them – there were always greater priorities for the resources at his disposal. The survey was suspended, although some preliminary reconnaissance for its completion was carried out by an engineer, M. Loire, in order to identify a set of mountain tops from Berry southwards that could form triangulation stations, visible from one another. His reconnaissance was not entirely successful, though, since the mountains that he identified centered along a line that drifted east of the Paris Meridian towards Bezier.

Louvois died in 1691 and Cassini sought permission and finance to restart the project. Louvois' successor, the Comte de Ponchartrain, agreed and fieldwork was resumed in 1700, starting from the tower of the Cathedral at Bourges (Cassini 1720). Cassini I was aided by his son Cassini II and Giacomo Maraldi.

People Giacomo Filippo Maraldi (1665-1729)

Giacomo Maraldi (his name is rendered into French as Jacques Philippe Maraldi and he is known as Maraldi I) was the son of Angela Cassini, sister of Cassini I, and therefore Cassini's nephew. He was called by his uncle to become his working colleague in Paris in 1687. He collaborated with Cassini on the timing of Jupiter's eclipses and defended Cassini's misdirected opposition to Rømer's view that the timing irregularities of Jupiter's satellites were caused by the time it took light to travel from Jupiter to Earth. Throughout his lifetime he worked on a catalogue of star positions, which was never published. Delambre (1827), no admirer of the Cassinis or anyone associated with them, caustically remarks that "we may infer that his uncle did not have a very lively interest in this fundamental branch of astronomy" and suggests that Maraldi's comments on the speed of light were more in the nature of a defence and an advocacy of Cassini's views than an objective search for truth.

The surveying expedition reached the border with Spain in 1701 and they erected a stone pillar at the southernmost extremity, on the 2780 meter summit of le Pic du Canigou in Rousillon (French Catalonia).

Places Canigou

This snow-topped mountain is one of the highest in the Pyrenees, and certainly one of the most imposing as it stands alone in the eastern end of the mountain range. It is regarded with awe by Catalonians as sacred, especially by those in the French part of Catalonia, who consider it a symbol of nationhood and its significance is proclaimed by bonfires lit at Canigou's summit at the summer solstice. The geodesic stone pillar erected by Cassini's team apparently does not presently exist and what can be seen from a distance at the peak is a cast iron cross erected by a team of boy scouts. (I suspect, without hard evidence, that they used the base prepared by the surveyors).

The work to map the southern part of the Paris meridian was again interrupted, this time by the War of the Spanish Succession (1701-1713). This war arose from the death of Charles II, king of Spain, who, childless, left his kingdom to his French grandson, the duke of Anjou who became Philip V and was also in line of succession for France. The strengthening of an alliance between Spain and France was threatening to other European powers, and ultimately this led to a war with France and Spain on one side, and Austria, the Netherlands, most of the German states and England on the other (smaller states, including Catalonia and Valencia, shifted from one side to the other as fortune ebbed and flowed or as the political situation was expedient).

The war started well for France in the Alsace but then France suffered defeats in Bavaria (at the Battle of Blenheim in 1704) and Belgium (at Ramillies in 1705 and at Oudenarde in 1708). At one point in 1709, Austrian and British troops moved from the north towards Paris but were defeated by France in the Battle of Malplaquet, near the border between France and Belgium. Shortly thereafter, Britain and France grew tired with fighting each other and Britain began to fear the successes of its ally, Austria. The war among France, Britain and the Netherlands was terminated by the Treaty of Utrecht in 1713 by which Philip V gave up his ambitions to the throne of France and France ceded possessions in Canada to Britain. Britain left the Catalans to fight for themselves and France besieged Barcelona, which fell in 1714, with Catalonia and Valencia coming under the control of King Philip V. Franco-Austrian hostilities were ended by the Treaty of Baden in 1714.

This was not a good ending for France, and the air of gloom at the finish of the war was deepened by the death of Louis XIV and other members of his family in 1715. His funeral was, however, a cause for the poorer people to get drunk in celebration because they had suffered so much from hunger in his lifetime. The end of his reign, by which he had lost almost all that he had earlier gained, contrasted with the optimism of his accession to the throne.

People Gabriel-Philippe de la Hire (1677–1719)

Gabriel-Philippe de la Hire (Philippe II) was the son of the astronomer Philippe de la Hire and was immersed in astronomy and mathematics from an early age, assisting his father in his observations. In 1702, however, he was involved in a dispute with Jean Le Fèvre, editor of the *Connaissance de Temps*, who accused both father and son of plagiarism. Le Fèvre was expelled from the Academy as a result of this incident. Although he did not produce anything innovative, La Hire worked, like his father, on a wide range of problems in the arts and sciences.

Almost all of Louis XIV's legitimate children had died in childhood and the only one to have a child died soon afterwards and predeceased the King. The succession was passed from Louis XIV directly to his grandson who came to the throne as Louis XV at the age of 5. His uncle, Philip II (Duc d'Orleans) was declared Regent. The terms drawn up by Louis XIV before he died to effect all this attempted to restrict the power of the Duc d'Orleans but eventually he took full control as Regent.

People Nicholas Louis Lacaille (1713–1762)

Abbé Lacaille is best remembered for his expedition in 1750–53 to the Cape of Good Hope where he measured the positions of the southern stars (nearly 10,000 of them, compared to the previous best catalogue by Edmund Halley in 1677–78 of 350) and discovered over forty nebulae and star clusters. He systematized and named fourteen southern constellations, including some "modern" names which now seem quaintly historic, such as Antlia, the Air Pump, Circinus, the Compasses, Fornax, the Chemical Furnace, Octans, the Octant, and Pyxis, the Mariner's Compass.

   Lacaille was born near Reims, the son of a *gendarme* and a member of an old and distinguished family. He developed an interest in mathematics and astronomy, and, self-taught, took up an appointment at the Paris Observatory under Cassini in 1737. He was assigned to help with the mapping of the Atlantic coast, where his work was so impressive he was chosen to help Cassini on the meridian measurements in 1739. Lacaille took the most energetic, leading role in the measurements measuring baselines at Bourges, Rodez and Arles and positions astronomically at Bourges, Rodez and Perpignan. Even in the severe winter weather of 1740/41 he surveyed across the mountains of the Auvergne and around Paris until the spring of 1741, re-surveying Picard's baseline at Juvisy which he found to be 0.1% too long.

With this background one can certainly understand why Lacaille would have replicated a geodesic survey in the Cape, where he measured the length over ¾ degree of latitude. He was puzzled to find his results suggested the Earth was prolate (pointed at the poles) not oblate (a flattened sphere). This seems to have been an error introduced into his measurements by the gravitational deflection by Table Mountain of the plumb-bob used to determine the vertical set-up of the instruments. Lacaille died at the comparatively early age of 49 of a fever brought on by the rigors of his observing regimen.

None of the political turmoil, uncertainty of authority, diversion of taxes to fight the wars, and the war-zone status of parts of the country was favorable to the

Fig. 14 The meridian of Paris was measured successively in a series of conjoined triangles that spanned France from north to south, from Dunquerque on the coast of the English Channel near the border with the Netherlands to Perpignan on the border with Spain. From the measurements of position and scale that had been made, the geodesists interpolated or extrapolated to fix the meridian line, which then acted as a fixed line to which the rest of France could be related. A second line was drawn east-west across France at the latitude of Paris from Brest on the coast of the Bay of Biscay to Strasbourg on the border with Germany, to provide a second reference axis. These two reference axes were surveyed repeatedly over the two centuries of the project to map France that was organized by the Paris Academy of Sciences. César-François Cassini de Thury (1744)

scientific mapping of France. Given the difficulties of the war period and the disorganization that followed the change of regime, it was some time before the mapping of France restarted. Cassini I had died in 1712 and it was not until 1718 that the survey resumed, completed by Cassini II, Gabriel-Philippe de la Hire and Maraldi I, who re-measured along the meridian to the north and completed its extension from Amiens to the northern coast at Dunkerque.

The north-south axis of the framework that could map France had been firmly established (Fig. 14). In 1733, to establish the second axis, Cassini II (aided by his son Cassini III) began measuring an arc at right angles to the meridian and running from Brest to Strasbourg. It was linked to yet another determination of the meridian carried out by Cassini de Thury (Cassini III), Nicholas Louis Lacaille and Giovanni Domenico Maraldi.

People Giovanni Domenico Maraldi (1709–1788)

Giovanni Maraldi (Jean Dominique Maraldi, Maraldi II) was the nephew of Maraldi I, brought by his uncle to Paris in 1726. He worked with Cassini II and Cassini III on a number of the large observing programs at the Paris Observatory for over 40 years before returning to Italy. On the death of Lacaille, Maraldi put his star charts through the press and wrote a preface on the life of his lifelong colleague and friend. In the opinion of Delambre (1827) Maraldi was "an industrious and worthy astronomer." Since Delambre did not have many enthusiastic things to say about the Cassinis or anyone connected with them, we may take this faint praise to imply a more positive assessment than at first it seems.

People Louis-Guillaume Le Monnier (1717–1799)

On the 1739 expedition Cassini also took along the younger brother of the astronomer Pierre Le Monnier. At the age of 22, Louis-Guillaume Le Monnier was able to make a natural history survey of the southern regions along the meridian, particularly the botany and the mines, and wrote a memoir, *Observations d'histoire naturelle faites dans les provinces méridionales de France, pendant l'année 1739* (1744) "containing the detail of all the plants and other natural curiosities that he found in his voyage, and which enrich the Garden of the King." He noted interesting ecological trends, and in his thirties he worked as a physicist on fluids and electricity. His work attracted vitriolic criticism from a rival, Abbé Nollet, which destroyed his self confidence and he wrote little in his later life. He had an apartment in the Tuileries and was home when a mob of Revolutionaries attacked it on August 10, 1792; he was saved by an unknown individual. According to historian L. Plantefol, "Destitute, he retired to Montreuil, where he lived a miserable existence on the income from an herbalist's shop which he opened."

THE ACCURATE SURVEY of Paris by triangulation made it possible to make a precise measurement of the speed of sound, which traveled so fast it was necessary to make the measurements over large distances. The Academy of Sciences addressed this question of physics repeatedly with the aid of the Prime Meridian survey, and the principle of the method was outlined by Marin Mersenne. Mersenne noted that it was well-known from observing battles that the cannon's flash precedes the sound of its loud explosion. The light moves so quickly that its speed cannot be measured, but the sound follows at an interval that Mersenne estimated by counting his heartbeats or pulse.

In 1640 Mersenne gave a figure of 1380 Parisian feet (*pied du roi*[13]) per second for the speed of sound (450 meters per second). There was a discrepancy between Mersennes' figure and that of Pierre Gassendi in 1635 (1473 Parisian feet per second) and Robert Boyle (equivalent to 1125 Parisian feet per second). Boyle explained this as something arising from a difference in the consistence of air in France and Britain. The most precise result, however, was achieved by a measurement by

---

[13] The *pied du roi* in Paris was the equivalent of the English foot of 12 inches. It measured 12.8 inches in the English system or 325 mm in modern metric units.

William Derham (1657–1735) in 1709 in Britain, which resulted in a speed-for-sound equivalent to 1072 Parisian feet per second.

The first experiments by the Academy of Sciences carried out by Cassini I, Huygens, Picard, and Romer produced the result that sound passes at a speed of 180 toises, or 1080 Parisian feet per second. The experiments were repeated in March 1738 by Cassini, Maraldi and Lacaille, exploiting the measurements of the survey; observation points were separated by large, accurately measured distances over which it took a considerable time for the sound of the cannon to travel. The chosen stations were the *mire* on Montmartre, the Paris Observatory, the mill at Fontenay aux Rose and the tower of Montlhéry, 29 kilometers from Paris. Many experiments were carried out with cannons of different sizes and with different charges of powder, day and night and in different weather conditions. These measurements gave an average of 173 toises (1038 Parisian feet) per second. The reduction of 34 Parisian feet per second between this value and the value by Derham would accumulate to a difference for the travel time of the sound of the cannon of 3 seconds between Montlhéry and Montmartre. Cassini estimated that a measurement error of half a second would be the most there could be, less than 1% of the two minutes it would take for sound to travel 29 km.

Places The Tour de Montlhéry

The Tower of Montlhéry is the main structure that remains of a small castle dating to the 14th century. It stands on an isolated hill directly south of Paris and has been occupied as a defensive position since AD 991. Its prominent high location within sight of and immediately south of Paris meant that it was a very convenient landmark for the trigonometric surveys of the Paris Meridian. However the castle was damaged during the period of the Religious Wars, and The Tower has recently been (controversially) restored. It stands high above the stone foundations of the castle buildings, surrounded by what may have been (on top of the steep hill) a dry moat; in an attractive grove of trees, some flowering prettily in the spring, and full of birds. Although it is a pretty woodland setting, it is not easy to imagine that during the eighteenth century measurements of the speed of sound were taken at this location, and the firing of a cannon north of Paris could be heard. Now, one hears the continuous sounds of motor vehicles on the motorways criss-crossing the plain below and the noises of airplanes taking off and landing at Orly airport a few kilometers away.

Cassini discovered that the speed of sound depended on wind direction, with or against the direction of sound, with the speed increasing or decreasing by the wind speed. He wanted to find out if the speed of sound depended on the season or the geographical location and aimed to carry out experiments during the expedition to southern France which he was about to embark on. Ultimately, though, he did not find it practical to spend time on the experiments as he had planned, as they were a distraction to the main purpose of his work, to survey France.

WHEN THE SURVEY of 1739 was completed, France had been surveyed north-south and east-west, and its extent had been precisely determined. The measurements were used as a series of starting points to construct a map of the whole of France (Robb 2007) and reached unprecedented accuracy. The surveying expeditions concentrated along river banks, partly for ease of access and partly because the resources along the river valleys were economically the most important. The surveying teams erected tall pyramids, or scaffolding, at 10 km intervals on the two axes to see over the trees lining the river. From these scaffolds, a surveyor could sight on prominent landmarks in the towns of France like the spires of churches.

The peasants became suspicious, this sentiment arising from fear and superstition; the surveyors used cabalistic instruments, stared at the stars, and wrote hieroglyphics on papers. Was this witchcraft, they wondered? This was justified concern that the investigations of these agents of the state would result in an extension of control and taxes. In further cases, even educated local officials were suspicious of these strangers. Were they spies, making signals and plotting revolt? During 1743 the teams were working in the Vosges, in an area known to be populated by Anabaptists, as noted darkly by a local official, M. de Blinglin. These people worked on Sundays and left people behind or paid peasants to light signal fires at specific times; one had left a stone marked J.G; another made copious notes. De Blinglin reported all this to his superiors, but their reply of reassurance did not completely satisfy him. Even if their mission was for some "official" purpose, perhaps they were also carrying out activities as part of a treasonable plot?

Some of the suspicions about the surveyors had not even this amount of rationality as their basis. In southern France near the village of Les Estables (Haute-Loire) in a remote region at the foot of the Mézenc mountain range, the arrival of a surveyor coincided with several items of bad luck. It had happened before when similar people had come from the capital – cows had become lame, sheep were found dead, crops withered. The man was attacked, bludgeoned and hacked to death by the villagers before he could work his malevolent sorcery. Attacked in the same way for similar reasons, another surveyor working near the village of Cuq in the Tarn was lucky to escape with his life. He recovered consciousness and staggered, bleeding and nursing a broken skull, to an inn run by the "Widow Julia", who called for help and a doctor. The surveyor recovered but had to retire from active work.

We do not know which particular one of the ninety five surveyors who worked on the project lost his life, a martyr to geodesy. Of most we know only their last names, enscribed on the charts into which their work was compiled. There are fragmentary details of a dozen. One had left the church and later died of a disease that he contracted in southern California where he went to observe the transit of Venus in 1769. One had been a playwright. One became a teacher, another became a general. The work that they created was a series of more than 800 accurately-surveyed triangles relating the principal cities and towns on the map of France. Under Cassini III, the surveyors and cartographers completed the most comprehensive map of France in 182 pages, even though the government withdrew financial support. With only two leaves of the map left to be printed, Cassini III died of smallpox, and the *Carte de France de Cassini* was published by his son, Cassini IV, in 1790.

Chapter 3
Shape of the Earth

Financed by the French government as a practical measure to improve knowledge about France, the work on the Paris Meridian became married by the Academy to an intellectual rather than practical scientific investigation – a test of theories of gravity. The applied "big science" project became a pure "big science" project. The project was a test to set two teams of scientists on adventures beyond the boundaries of France, north into the icy wastes of Lapland and south to the equator in present-day Ecuador, its aim to measure the shape of the Earth.

Since antiquity, educated people knew that the Earth was approximately spherical; the controversy as to whether Christopher Columbus would fall off the edge of the world if he sailed from Spain westwards was founded in ignorance. The real doubts centered on whether he would survive the dangers (weather, sailing hazards, sea-monsters, etc.) and whether he would be able to achieve the objective, namely an alternative route to the East Indies. As history tells us, he thought that he had indeed achieved his objective when he found what he called the West Indies, which we recognize now as part of the Americas.

The well-rehearsed arguments about the sphericity of the Earth included the fact that the shadow of the Earth cast on the Moon during a lunar eclipse was always circular and therefore must be the shadow of a spherical body (Fig. 15); it was well known, too, that a lookout at the top of a ship's mast would see the last glimpse of land long after his fellow sailors on deck had lost sight of it (Fig. 16).

The size of the Earth was also known to a good approximation, as well as its shape. In the third century BCE, Eratosthenes (c. 276-c.195 BCE), the librarian of Alexandria, determined the size of the Earth; he had heard that at Syene in Upper Egypt (present day Aswan) the Sun was directly overhead at noon on the day of the summer solstice. The Sun's rays reached the bottom of a deep well, and he determined the length of the shadow of a vertical post at Alexandria on the same day and found that the angle of the Sun was 1/50 of a circle to the south of the zenith. He believed that the distance between the two cities was 5,000 stadia, and it is suggested he determined this by driving a carriage between the two cities and counting the revolutions of the wheels. Thus, the circumference of the Earth was 50 times this, or 250,000 stadia. The equivalent of this length of stadia is not well known but is believed to be about 45,000 kilometers, remarkably close to the modern value of 40,000 km for the circumference of the Earth.

P. Murdin, *Full Meridian of Glory*,
DOI: 10.1007/978-0-387-75534-2_3, © Springer Science + Business Media, LLC 2009

Fig. 15 The Earth's shadow was revealed as the Moon passed through its lower edge during the total lunar eclipse of November 8, 2003. This photo is a composite of five images taken 40 minutes apart, centered on the Earth's shadow. The middle image, taken at the middle of the total eclipse (and therefore exposed more than the others), shows the color of the Moon in the Earth's shadow. The shadow is not totally dark because the Earth's atmosphere acts as a lens and bends light into it and the light is tinted red as it grazes the Earth's atmosphere, just as in sunsets. The edge of the Earth's shadow (more precisely, the edge of the umbra) is marked in the image. It is circular, and the Earth must be spherical. © 2004 Tom Matheson

Fig. 16 A lookout on shore would be able to see the top of a ship's mast as a ship left the land, long after the hull and sails of the ship had disappeared below the horizon, hidden below the curvature of the Earth.

In the seventeenth century, it had become clear that the Earth was not entirely accurately spherical. The evidence proving this was the variability in the length of the pendulum required in a pendulum clock to beat seconds.

THE CONCEPT OF THE pendulum clock was due to Galileo. According to his first biographer Vincenzo Viviani, while a student at the University of Pisa Galileo began his study of pendulums after watching a suspended lamp swing back and forth in the cathedral (presumably during a boring sermon.) He timed the period of the swing with his pulse beats, beginning serious investigations in 1602 and discovering that the period of a pendulum's swing is independent of its amplitude – in other words, the time that the pendulum takes to traverse the arc of the swing is always the same no matter how long the arc is. The isochronism of the pendulum meant that it had an immediate application as a timepiece, and in 1603 a friend of Galileo's, a physician in Venice named Santorio Santorio, began using a short pendulum, which he called "pulsilogium," to measure the pulse of patients. It was nearly 40 years later that Galileo realized how the pendulum could be used to improve a mechanical clock.

Mechanical clocks, using a heavy weight or a spring to provide power, had replaced hour-glasses or water clocks (which measured time through the regular flow of sand or drip of water). They worked by using an escapement, a lever that pivoted and meshed with a toothed wheel at certain intervals. They were not very accurate, though. They could lose or gain approximately 15 minutes per day. In 1641, at the age of 77 and completely blind, Galileo realized that a pendulum could be connected to the escapement to regulate it. In 1658 in his biography of Galileo Viviani described what happened as follows:

> One day in 1641, while I was living with him at his villa in Arcetri, I remember that the idea occurred to him that the pendulum could be adapted to clocks with weights or springs, serving in place of the usual tempo, he hoping that the very even and natural motions of the pendulum would correct all the defects in the art of clocks. But because his being deprived of sight prevented his making drawings and models to the desired effect, and his son Vincenzio coming one day from Florence to Arcetri, Galileo told him his idea and several discussions followed. Finally they decided on a scheme … to be put in practice to learn the fact of those difficulties in machines which are usually not foreseen in simple theorizing.

Galileo never made the pendulum clock that he had conceived, and it was Christiaan Huygens who first invented the pendulum clock in about 1656. He patented it in Holland, but his patent application in Italy was refused on the grounds that he had plagiarized Galileo. His patent application in France was also refused three times on the more pragmatic grounds that the master clock-makers of Paris might object. He published his work on the design and the science of pendulum clocks (*Horologium Oscillatorum*) in 1665, finding that he could readily achieve an accuracy of 10 to 15 seconds per day, an enormous advance over the mechanical clocks available until that time.

Within a year of publishing *Horologium Oscillatorum*, Huygens was invited by Colbert to become a member of the Paris Academy of Sciences. At the same time he set about developing the design of the pendulum clock and investigating its scientific potential and found as part of the regulation of the clocks it was necessary to determine the length of the pendulum that gave a calibrated time scale– the seconds pendulum, namely one whose swing takes one second from end to end (two seconds for the period).

People Christiaan Huygens (1629-1695)

Born in The Hague in the Netherlands, Huygens studied in Leiden and Breda. His father corresponded with Marin Mersenne and his son eventually joined in the correspondence; Mersenne challenged Huygens with mathematical problems and stimulated his interest. Huygens took up telescope-making and with one of them discovered the large satellite of Saturn, Titan, and the true nature of the planet's ring system. Although his health was never robust, he was widely traveled and visited Copenhagen, London and Paris. In 1666 he joined the Academy of Sciences at its foundation helping to set it up along the lines of the Royal Society of London, of which he became a fellow in 1663. He kept his association with the Academy, even when France was at war with the Netherlands during the War of the Spanish Succession, until 1681. In 1689 he met Isaac Newton in London and in 1698 published (posthumously) one of the first books about the possibility of extraterrestrial life (over the next 15 years it was translated into English, French, German and Russian, a rapidity that would be remarkable even today). His association with Gian Domenico Cassini is remembered in the naming of the HUYGENS space probe that was landed on Saturn's moon, Titan in January 2005, carried there by the CASSINI spacecraft.

Other scientists were working to measure the length of the seconds pendulum, but different observers reported remarkably different values. One cause of the variation in the length of the seconds pendulum was the temperature of the clock. The rate of a pendulum depends on its length, therefore a given pendulum changes its beat as a change of temperature causes its length to alter. The temperature rises, the metal of the pendulum wire or rod expands, the pendulum length increases, and the pendulum oscillates slower.

People Marin Mersenne (1588–1648)

Marin Mersenne was a member of the order of the Minims who taught philosophy in Nevers and moved to Paris in 1619. A mathematician, he is known particularly for his discovery of the *Mersenne Prime Numbers*. Prime numbers are numbers that cannot be subdivided into factors other than themselves and 1. Mersenne observed that a surprising number of prime numbers are of the form $2^n - 1$. 3 is prime (it can only be factored as 3×1) and it can be written $2^2 - 1$. 7 is prime (7×1) and can be written $2^3 - 1$. 31 is prime and can be written $2^5 - 1$. And so on. There are 44 Mersenne prime numbers known (as of 2008). $2^{32,582,657}-1$ is the largest, discovered September 2006. It has over 9 million decimal digits and is tantalizingly close to winning the reward of $100,000 that has been offered for the first prime number discovered with over 10 million digits.

Mersenne is also famous for his position at the center of a network of scientific discussion. His cell in the monastery of the Minims de l'Annonciade near the Place Royale (the present-day Place des Voges, the oldest square that survives in Paris), was a meeting place for scientists and mathematicians like Fermat, Pascal, Gassendi, and others. These meetings and people developed into the Academy of Sciences when it was founded by Colbert. After Marsenne's death, letters were found in his cell from 78 scientists; he disseminated results from one to the other in a web of correspondence that was the precursor of today's scientific journals and preprint archives on the Internet.

Huygens did not believe in this imperfection of his clock. Describing the pendulum clock in his book *Horologium* (1658), Huygens referenced the Belgian astronomer Godefroy Wendelin as observing that a pendulum clock beat faster in winter than in summer. Huygens dismissed the observation as a mistake by Wendelin in calibrating the pendulum clock against hourglasses and sundials– they were too inaccurate to make the comparison, said Huygens. Even as late as 1690, Huygens could not believe Denis Papin's view that pendulum rods expand if they are taken from cooler to warmer climates, in spite of the evidence to the contrary supplied by Picard and La Hire. Picard observed that an iron rod, which in winter during freezing weather was 1 foot long, was 1 foot and ¼ ligne[14] long when heated by a fire. La Hire observed that an iron rod which was 6 feet long in the winter, was 6 feet and 2/3 ligne long in the summer.

In 1671 the Academy of Sciences sent Jean Richer to Cayenne, Guyana, South America, on the equator, to observe the close approach of Mars to the Earth in 1672. He was instructed to observe its position to establish its distance and thus the scale of the solar system, but to do so he needed an accurate clock. He took a pendulum that had beaten seconds in Paris and in Cayenne re-regulated the clock against the motion of the Sun and stars. Brought straight from Paris and set up again unaltered, off the boat, the pendulum ran slow and lost two and a half minutes every day. To make it beat seconds in Cayenne, Richer had to shorten it by 1¼ gnes (about 3 millimeters). Richer did a good job to establish the credibility of these measurements by repeating them weekly over nearly a year throughout an annual temperature cycle. This gave him evidence that the disparity was not due to the temperature difference between tropical Guyana and temperate France because the discrepancy was too large.

People Isaac Newton (1643-1727)

Sir Isaac Newton (Fig. 17) was born into a Lincolnshire farming family and when, through lack of interest, he made a mess of managing the estate, he was sent to Trinity College, Cambridge University to study, becoming a fellow there in 1667. At first he was interested in pure mathematics and developed what we now know as the calculus. (He carried on an acrimonious dispute with Gottfried Leibnitz over who had priority for this discovery, Leibnitz having discovered the same multiplication independently.) He turned to optics and demonstrated how white light was made up of rays of light of different colors that it refracted differently through a prism and formed a spectrum, ultimately publishing his theories in *Optiks*. In addition, he invented and constructed a reflecting telescope and developed laws of motion that explained the motion of bodies (such as balls that impacted or whirled on a string). Turning his attention to the motion of a planet around the Sun, he conceived of a force of gravity that acted like string that tied them together and showed that if the force behaved as a function of distance according to an inverse square law it would give rise to Kepler's laws. He published his results in a book that became hailed as a scientific masterpiece, the *Principia*.

In 1699 he moved to London and became Master of the Mint. He spent his later years in religious and astrological investigations of little scientific interest.

IN HIS BOOK *Principia* (Proposition 20 of Book 3), Isaac Newton quoted Richer's results on the period of a pendulum on the equator and in France. He did so with approval, commending the "diligence and caution," which "seem to have been lacking in other observers." Newton offered a solution to Richer's discovery that a pendulum beat slower at the equator than in France; he interpreted his results as being due to the shape of the Earth, which bulged at the equator it was rotating and was correspondingly squashed at the poles. As a result, gravity was reduced at the equator and the pendulum beat more slowly.

---

[14] 1 French *ligne* = 0.09 inch, or 2.25 mm.

Fig. 17 Isaac Newton as a young man, dressed in the robes of a bachelor of arts, as he appeared at the height of his scientific creativity

People Jean Richer (1630-1696)

Little is known of Richer's life except for his published work. What is known is that he became a member of the Academy in 1666 and carried out various tasks for the Academy in France, Canada and Cayenne. He observed the opposition of Mars in 1672 and established its parallax or distance, which determined the most accurately known value of the scale of the solar system (the astronomical unit) of the time. When he returned to Paris he transferred to work on military engineering problems, such as the design of fortifications – it is not known what provoked this change of activities.

Newton gave an explanation for his results based on his theory of gravity and his laws of motion. He did so with a powerful intuition that others did not find easy to follow (Greenberg 1995). Johann Bernouilli, the third most able mathematician of the time (after Newton and Gottfried Leibnitz), confessed: "I tried to understand it. I read and reread what he had to say concerning the subject, but … I could not understand a thing. I do not know whether the fault lies with my impatience, resulting from my reaction to references to things back in Book I or whether I do not understand how he applied these things [to the explanation in Book III of the *Principia*]. In a word, all I found was gibberish and obscurity."

Ideas The Principia

The seeds of Newton's masterpiece, the *Principia*, were planted in 1684 at the Royal Society. Two astronomers, Edmund Halley and Sir Christopher Wren, suspected from Kepler's Third Law that there was an inverse square force governing planetary motions but could not prove it. During a discussion of this point Robert Hooke claimed that he could prove all of Kepler's laws. When Wren cast doubt on the claim and offered a book as a prize for a public proof, Hooke failed to deliver.

Halley put the question next to Newton. Newton had already solved the problem years before but could not find the proof among his papers. Later he sent Halley a nine page proof, *De Motu Corporum*, or *On the Motions of Bodies in Orbit*. Halley suggested publication, but Newton was reluctant to expose his work to public scrutiny, having no appetite for the cut-and-thrust of debate. Halley persisted and Newton worked for 18 months on the paper, expanding it to three books where the mathematical theories of gravity and motion are set out like a book of geometry with theorems and propositions.

The work was published at Halley's expense as *Philosophiæ Naturalis Principia Mathematica*. It was written in Latin to make it accessible to scientists throughout the world. Its English title is *The Mathematical Principles of Natural Philosophy*, but it is always known by its short title of *The Principia*, pronounced "*Prinsipia*" or "*Prinkipia*." It was first published in 1687, followed by a second and a third edition in 1713 and 1726 respectively; copies of the successive Latin editions were immediately available in France. It was first translated into English by Andrew Motte (1729). A French translation was prepared by the Marquise de Châtelet with explanations and reworkings of the theorems by Alexis-Claude Clairaut (1756).

For the purposes of his explanation of Richer's results, which Bernouilli found obscure, Newton envisaged that the Earth was bored through with a tube that ran from its pole along a radius to its center, and then from its center along a radius at right angles to the first, to its equator (Cohen and Whitmore 1999). The tube was filled with water, and the weight of the water along the polar axis was equal to the weight of the water to the equator because the water tube was connected at the Earth's center and the pressures of the two columns must match at the central point (Fig. 18). But the weight of the tube of water at the equator was reduced by centrifugal force because of the rotation of the Earth; for the weights to be equal, the tube to the equator had to be filled with more water than the tube to the pole. Newton calculated that the equatorial radius of the Earth was longer than the polar radius by about 17 miles, and the ratio of the diameter of the Earth at its equator to its diameter at the poles is 230 to 229 miles.

There is much in this proof that is unstated and others besides Bernouilli have found it subtle to understand, including the Nobel prize-winner Subramanyan Chandrasekhar (1910-1995) in his reworking of the *Principia* in modern mathematics (1995).

To Newton the result that the Earth was flattened at the poles was perfectly plausible- Jupiter was so flattened at the poles that its elliptical shape was plain to see in a telescope. Newton knew that in 1691 Cassini I had measured Jupiter's diameter and its diameter at its equator was about 7% longer than its diameter at the poles; this was confirmed by James Pound's observations in 1717.

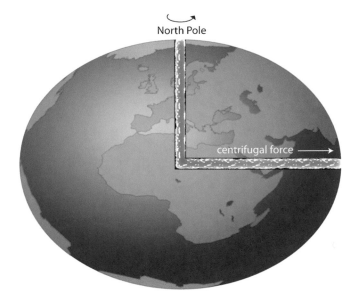

Fig. 18 If the Earth was drilled through with a tube in the shape of an elbow, filled with water and with its corner at the center, the pressure of water along each arm of the tube would have to be equal so that the pressure at the center was the same in both arms. Because the Earth is rotating around the North Pole, the weight of water in the equatorial arm is reduced by centrifugal force, so there has to be a greater length of tube to make the same pressure. Therefore the Earth is flattened at the poles

People James Pound (1669-1724)

Educated in the field of medicine, Pound took holy orders and was sent by the East India Company to be chaplain in Madras and then Chushan Dao, an island near Shanghai. He made astronomical observations while he traveled and sent them to John Flamsteed, the astronomer royal. Flamsteed gave him a quadrant with which to measure the positions of the southern stars, but unfortuneately it took four years to reach him and corroded on the way. In 1702 Pound moved to Pulo Condore, an island in the Mekong delta, and when in 1705 native troops attacked the settlement and set fire to the buildings, Pound and ten other Europeans escaped with just their lives.

Pound returned to England via Batavia and took up a quieter life as a vicar in Wanstead. He used a 15-foot telescope in 1716 and 1717 to make various planetary observations and in 1717 installed a 123-foot focal length lens that Huygens had made as the collecting lens of an aerial telescope (one with no tube), supported on a maypole removed from a London street by Isaac Newton. Newton used Pound's measurements of Jupiter, Saturn and their satellites in the third edition of his *Principia*.

Pound went on to support the interest of his nephew James Bradley (1692-1762) in astronomy and to work with him in discovering the phenomenon of the aberration of light.

For Newton the shape of the Earth was important since it not only in itself provided a test of his theory of gravity but also affected his calculations of the motion of the Moon and other applications of the theory. Newton therefore examined whether the latest measurements of the Earth could provide confirmation of its flattened shape more directly than his interpretation of measurements of the length of a seconds pendulum.

The Paris scientists had surveyed France both by geodesy and by astronomy. This not only accurately positioned parts of France relative to one another but also gave the scale of the map of France. It was a matter of calculation to relate the scale of the map from latitude to latitude, determining the length of a degree of latitude along the meridian. The length of a degree of latitude changes from the equator to the pole, according to the shape of the Earth. If the Earth is prolate (pointed at the poles, like a lemon, or a rugby or American football) a degree is longer at the equator than at the poles (Fig. 19). If the Earth is oblate (flattened at the poles, like a tangerine), the contrary is the case (Fig. 20).

In the *Principia*, Newton collected the available results from the French observers. They provided what seemed to be a good average value for the length of a degree. In the third edition of *Principia,* Newton quotes individual results as 57,060 toises from Picard's measurements between Amiens and Malvoisine, and 57,061 toises from the results of the two Cassinis from Collioure in the south of France to Dunkerque in the north. Newton noted that the mathematician and surveyor Richard Norwood (1590–1675) had measured the distance and the latitude difference between London and York and had found that the average length of a degree along this meridian appeared longer at the equivalent of 57,300 toises, encouragingly larger than the values in France further to the south. The French survey was undoubtedly more accurate than Norwood's, but Newton makes nothing of this apparent confirmation of his calculation. Voltaire gives an account of this part of Newton's work in *Letters from England* (1731):

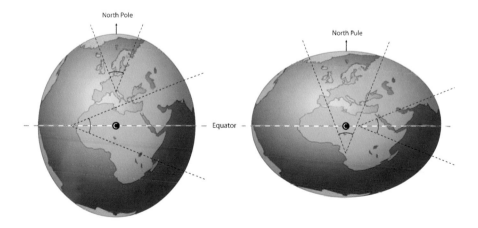

Figs. 19 (Left) and 20 (Right)   The length of an arc of the Earth's surface corresponding to a given angle of latitude varies over the Earth in a way that depends on the shape of the Earth. (Left) 1° of latitude is longer at the equator than at the poles if the Earth is prolate (pointed at the poles), and (Right) vice versa if it is oblate (flattened at the poles)

[A]t that time the English had but a very imperfect measure of our globe, and depended on the uncertain supposition of mariners, who computed a degree to contain but sixty English miles, whereas it consists in reality of near seventy. As this false computation did not agree with the conclusions which Sir Isaac intended to draw from them, he laid aside this pursuit. A half-learned philosopher, remarkable only for his vanity, would have made the measure of the Earth agree, anyhow, with his system. Sir Isaac, however, chose rather to quit the researches he was then engaged in. But after M. Picard had measured the Earth exactly, by tracing that meridian which redounds so much to the honor of the French, Sir Isaac Newton resumed his former reflections, and found his account in M. Picard's calculation.

Picard had established the size of the Earth but was not able to calculate the shape of the Earth. The variation across France of the length of a degree of latitude was too small even for the most accurate survey to distinguish the measurement errors. Newton calculated what he would have expected:

The change in the length of one degree latitude over about ten degrees latitude between southern France and Britain (say from 40° to 50°) was only some 100 toises in 57,000, or about 0.02% per degree of latitude. The accuracy required to measure such a small change was challenging. Let us see how difficult it is to distinguish whether the Earth was flattened at the poles or not by this method. Suppose that a quadrant can measure angle to 1 arc minute. If a surveyor triangulates from each end of a perfectly accurately measured baseline towards a landmark that is one degree of latitude away, or 110 km, he can determine the position of the landmark to 30 meters, or 15 toises. That is about the size of the difference in the length of a degree of latitude from one degree to the next and it would be tough to be certain of such a small difference; to be confident of the result it would be necessary to use a surveying instrument accurate to arc seconds. This technical development was decades away.

Newton's calculation of the shape of the Earth

Latitude of the place	Measure of one degree on the meridian	Latitude of the place	Measure of one degree on the meridian
degrees	toises	degrees	toises
0	56,637	45	57,010
5	56,642	6	57,022
10	56,659	7	57,035
15	56,687	8	57,048
20	56,724	9	57,061
25	56,769	50	57,074
30	56,823	55	57,137
35	56,882	60	57,196
40	56,945	65	57,250
1	56,958	70	57,295
2	56,971	75	57,332
3	56,984	80	57,360
4	56,997	85	57,377
		90	57,382

NEWTON'S WORK ON GRAVITY was published in the first edition of the *Principia* in 1687 and by 1701 Cassini I had finished the survey from Paris to Bourges. Through his reckoning, the scale of a degree was 57,097 toise, compared to the value of 56,996 toise that he quotes as measured by Picard from Paris to Amiens. His conclusion was that each degree was 1/800 (0.1%) shorter as you approached the pole. This was five-times bigger than Newton had calculated and in the opposite direction implying that the Earth was prolate, or pointed at the poles. His son, Cassini II, confirmed his father's conclusion that the Earth was pointed at the poles in 1718 when he had finished the measurement of the Paris Meridian from Perpignan to Dunkerque: "The success of this work gave us room to conjecture that the degrees of the meridian increase as they approach the equator," wrote Jacques Cassini in 1718, and declared in a presentation to the Academy that the Earth was a prolate spheroid, the shape, as some had likened, of a pot-bellied man wearing a belt. This most certainly brought chuckles from the well-fed Academicians in the audience.

This observational result was theoretically backed in 1720-22 by a mathematician, Jean-Jacques D'Ortous de Mairan. Accepting the pointy Earth as a fact, Mairan concluded that Newton had obtained the wrong answer from an incorrect theory of gravity. What ought to be done, argued Mairan, was to use the geodetic observations showing the shape of the Earth to derive the law of gravity. Mairan suggested that the force of gravity at the Earth's surface depended on the radius of the surface's curvature. If you fitted a sphere inside the pointy poles of the prolate shape of the Earth that Cassini had measured, then the sphere would have to have a radius smaller than the Earth does on average. But the force of gravity was bigger. Conversely, if you fitted a sphere to the equator of the Cassini shape, you could fit a sphere the size of the Earth in one direction, around the equator, but the sphere would have a larger radius to be able to fit along the meridian and here the force of gravity would be weaker. Mairan conjectured that the force of gravity at the Earth's surface depends on the inverse of the product of the two radii of curvature at the surface. It was a logical possibility, but it was an arbitrary theory of gravity that fitted some results at the Earth's surface, and had nothing to do with any other property of gravity (for example the motion of the planets).

People Jean-Jacques D'Ortous de Mairan (1678-1771)

A physicist, Mairan worked in a Cartesian framework on theories of heat, light, the aurora, and gravity. He came from a noble family and had a wide interest in cultural activities – Chinese culture and the piano, for example.

In his essay, Mairan linked the nature of the force of gravity with the theory of gravity due to René Descartes, and this established a popular belief of the time that the pointy shape of the Earth was the result of the Cartesian theory of gravity. In the *Principia*, by now well known in France, Newton had clearly linked his theory of gravity with the prediction that the Earth was oblate rather than prolate. The shape of the Earth came to be seen as a test for one theory of gravity or the other. The two theories have been thought, at that time and recently as well, to differ in this practical way – if you measured the shape of the Earth you could determine which theory of gravity was right. As shown by John Greenberg (1995) and other historians of science, this widely held view was not true – the Cartesian theory has little to say about the shape of the Earth.

The two theories were very different in their basic philosophy. Newton's theory of gravity contained the remarkable concept of "action at a distance" for the force of gravity. Newton made the bold step linking the force of gravity at the surface of the Earth with the force that holds the planets in orbit. According to legend, he saw an apple fall from a tree and suddenly realized that this force must propagate into space to hold the Moon in orbit around the Earth. Even if the legend is untrue (it appears in print for the first time long after it was supposed to have happened), something must have triggered this thought in Newton's mind: the force that causes a pendulum to oscillate is the force that dominates in the solar system. It was as extraordinary to think of a force that extended from the Earth to an apple as one that reached to the Moon and beyond, meaning that gravity was "action at a distance."

It was logical, however, to think that the effect of the force would diminish with distance, and Newton postulated that the force of gravity between any two bodies varies as the inverse square of the distance between them. If two bodies are moved twice as far apart, the force of gravity between them diminishes by a factor four; if moved three times further the force diminishes by a factor nine.

With the aid of this law of gravity, and his other laws of motion, Newton was able to derive Kepler's laws about the motion of the planets. Kepler had discovered his laws by years of trial and error: Newton was able to derive them theoretically from more fundamental principles. This was a brilliant success that linked together seemingly quite separate issues – Kepler's laws, pendulum oscillations and the shape of the Earth – into one, unifying framework of calculations.

Philosophically Newton's theory of gravity left something to be desired. In particular, how could a push or a pull propagate across empty space, enabling one body to affect another? What carried the force? The alternative theory of gravity due to René Descartes had more appeal to common sense.

People René Descartes (1596-1650)

Descartes was a French philosopher who settled in Holland and published in both mathematics and mathematical philosophy (science), including meteorology, optics and astronomy. Since he was philosophically unconvinced by the notion of "action at a distance," or the Newtonian concept of the force of gravity, he developed a theory of vortices in the Universe and this theory carried the planets on their motions around the Sun. He is sometimes described as the first modern philosopher, the common x-y axes of a graph at right angles to each other, "Cartesian coordinates," are named after him, and "Cartesian logic" is the application of ruthlessly logical and consistent principles to an analysis.

DESCARTES ENVISIONED the Universe to be full of a material called the *plenum*. There was no space between planets and stars so therefore no vacuum could exist. The plenum had been set in motion at the Creation. If there were no empty spaces, how did matter move? Each body, reasoned Descartes, moved instantaneously into any space that had been vacated by a contiguous body. A body pulled or pushed another through a chain of motions of plenum material lying between them. The result was that the plenum rotated in vortices and acted on material bodies pushing them by friction. The Sun was embedded in a large vortex that acted by impulse on the planets, carrying them in orbit; each planet is at the center of a smaller vortex, transferring the rotation of the planet to any of its moons.

This concept of the way that the Sun and planets work produced a remarkably prescient picture of the formation of the Sun that we might recognize as very modern. If the Universe contained an immense number of vortices made up of particles of various sizes, in time they would make dust by rubbing against one another and would gather at the center of the vortex. This accumulation would become a star like our Sun; other vortices would settle into planets. Planets that move from vortex to vortex were recognizable as comets. All this is like the modern picture of the formation of the Sun and planets from a rotating cloud of dust and gas.

The Cartesian theory of gravity could not produce Kepler's laws from the vortex theory so it appeared second-best when compared to Newtonian theory in this way. It was difficult to find a natural Cartesian formulation that predicted the shape of the Earth (although the Academy awarded a prize to Bernouilli in 1734 for a paper in which he presented a set of equations showing how the swirling whirlpools would elongate the poles), and it is interesting that the fundamental philosophical difficulty with "action at a distance" remained a problem for centuries after Newton proposed it. In a way, though, the problem has been resolved by the General Theory of Relativity, which talks of space being bent by gravity: space is more curved near a body and this creates an orbit for bodies to follow as they pass by. The General Theory, therefore, approaches towards the Cartesian theory of a plenum.

Voltaire explained the battle between the two theories of Newton and Descartes in his *Letters from England* (1731):

A Frenchman who arrives in London will find philosophy, like everything else, very much changed there. He had left the world a plenum, and he now finds it a vacuum. In Paris the Universe is seen composed of vortices of subtle matter; but nothing like it is seen in London. In France, it is the pressure of the moon that causes the tides; but in England it is the sea that gravitates towards the moon; so that when you think that the moon should make the tide flood with us, those gentlemen in England fancy it should be ebb. In the Cartesian camp, everything acts by an impulse about which we understand not much; for M. Newton it is all done by an attraction about which we understand not much more. In Paris, you think of the Earth as a melon; in London it is flattened on both sides.

The last sentences refer to the shape of the Earth as delineated in the height of the tides, with the pointy Earth shape providing a tide opposite to the theory of the oblate Earth.

People Voltaire (1694-1778)

Francois Marie Arouet, who called himself Voltaire, came to notoriety in Paris as a young poet, writer and satirist. He enraged numerous enemies, suffering with dire consequences (including imprisonment and beatings). In 1725, he went to London to escape the threats. There he studied the differences between English and French sciences and philosophy, especially political philosophy. Returning to Paris he wrote prolifically, moving to Prussia and then Switzerland. He fled from city to city as the bitterness of his satire offended the church (he was a militant atheist) and other authorities, as he took up causes against them. At the age of 83 he was allowed to return to Paris, but after seeing the performance of his final play, he died in 1778.

The Cassinis' calculations about the shape of the Earth in 1701 and 1722 did not change Newton's mind after the first edition of *Principia* had been published. In the second and third editions he stuck to his guns; the pendulum observations seemed more conclusive than the geodetic measurements. He cited observations of the length of the seconds pendulum that had been made in places widely dispersed across the globe: St Helena, Guadeloupe, Paris, Cayenne, Martinique, Lisbon, Paraìba, Grenada, St Kitts, Santo Domingo, and Portobello. By contrast, the geodetic measurements all had to be taken within the confines of France, with the changes of scale from its south to its north being too small to determine accurately.

VOLTAIRE DESCRIBED the battle between the Newtonian view of the shape of the Earth and the reason that underpinned it, namely the theory of gravity, versus the geodesic measurement and the Cartesian theory of gravity (insofar as it was part of the argument). He cast the battle in national terms, pegged as a fight between Britain and France. Others saw it in the same way; for example, the secretary of the Academy, Bernard Le Bovier de Fontenele, rhetorically demanded to know why some French people wanted "to justify the English at the expense of the French? Who would ever had thought it necessary to pray to Heaven to preserve Frenchmen from a too favorable bias for an incomprehensible system, they who have so dearly, and for a System originating in a foreign land, they who have been charged with loving only that which is their own?" This view of the intellectual argument has been followed ever since.

However, this was not wholly a battle between England and France. In France, Newton's work was known and much admired, and his theory of gravity was defended against Mairan by French astronomers Joseph-Nicholas Delisle, Alexis-Claude Clairaut, and, leading the attack, Pierre Maupertuis.

People Alexis-Claude Clairaut (1713-1765)

Like his father, Clairaut was a mathematician, a child prodigy who read his first paper to the Academy of Sciences at the age of 12 and became a member at 18. He was a friend of Maupertuis and joined him in his belligerent advocacy of Newton's theories; they were two of a kind, both vivacious and attractive to women. After the expedition to Lapland, Clairaut published the theoretical interpretation of the measurements in *Théorie de la Figure de la Terre* (1743), considered a scientific classic. He developed the theory of the orbit of the Moon around the Earth, a technically difficult work of Newtonian mathematics because of the gravitational attraction of the Sun and the non-spherical shape of the Earth. In developing a theory to solve the so-called three-body problem of the mutual gravitational interaction of three bodies, he applied it to the reappearance of Halley's Comet, uncertain in its orbit around the Sun because of the attraction of Jupiter (and Saturn). His predictions of its reappearance reduced the error from a year to two weeks or less.

In the 1720's, the French mathematician Pierre Maupertuis was systematically building up for himself a reputation as a mathematician (Terrall 2002). His interests were wide and he cultivated an extensive network of social contacts among the French academic society elite. In the summer of 1728, he visited London for twelve weeks to broaden his network of influential colleagues among the Newtonian scientists of the Royal Society; Maupertuis undoubtedly discussed Newton's theory of gravity with the English scientists and certainly became familiar with the finely crafted instruments they used to make astronomical measurements.

On returning to France Maupertuis continued to investigate problems in geometry. Working with the Swiss mathematician Johann Bernouilli (1667-1748), he became interested in the shape of the Earth and attempted more rigorous and clearer mathematical proofs of Newton's explanations of its flattening. He was only partially successful and came to the conclusion that this problem could only be tackled by more accurate geodesy. In 1729 the Paduan scholar and polymath Giovanni Poleni (1683-1761) suggested that the cause of the discrepancy between Cassini's measurements and Newtonian theory was in the accuracy of the measurements, and in 1733 Poleni was awarded a prize by the Academy of Sciences. In accepting the prize, he presented a booklet on the topic to the Academy and simultaneously a commentary appeared in Holland on Poleni's results. Maupertuis followed up these discussions, reporting them to the meetings of the Academy.

The debate in the Academy was hard fought, with the iconoclastic, young Maupertuis and Clairaut lined up against a powerful establishment headed by the Cassinis. They scorned Maupertuis and dubbed him "Sir Isaac Maupertuis."

People Pierre Louis Moreau de Maupertuis (1698-1759)

Pierre Maupertuis (Fig. 21) was a charming man, unusually small, possessing gravity and austerity whilst also capable of being gay and passionate. Evidently he was attractive to women, although a difficult man to live with. At first he studied music but converted to mathematics and combined the two interests in his dissertation called "On the form of musical instruments." He became an enthusiastic proponent of Newtonian physics and spent ten years on measurements of the shape of the Earth in order to prove Newtonian theory.

In 1738 he visited Voltaire in Prussia and became a member of the Berlin Academy, but then in 1741 he was caught up in the war between Prussia and Austria and was taken prisoner at the Battle of Mollwitz. Luckily he was recognized by an officer and released, taking up residence in Berlin in 1745. He is known for proposing the *Principle of Least Action* which gave a means to calculate the route taken by a body in orbit, or a light ray through a lens. Maupertuis attracted many prominent scientists to Berlin and published some important works on embryology and genetics, but everything soon went wrong in the "affaire König", described by E. A. Fellmann in the *Dictionary of Scientific Biography* as the ugliest of all scientific disputes. Voltaire was brought into the affair and charged Maupertuis with plagiarism, error, persecution, and tyranny. He made fun of Maupertuis' expedition to Lapland and his amorous proclivities, satirizing him in a short story entitled *Micromegas* (1752). This story, regarded as one of the first works of science fiction and reminiscent of Jonathan Swift's *Gulliver's Travels* (1726), concerns an enormous visitor from a planet of Sirius who ridicules the small size and small minds of people on Earth, especially an expedition of astronomers (Maupertuis and his expedition, with the two sisters, returning from Lapland). Maupertuis was crushed, withdrew from Berlin, and in ill-health made his way towards home but unfortunately died in Switzerland before he could reach it.

People Samuel König (1712-57)

Samuel König, a Swiss scientist and mathematician, was a protégé of Maupertuis', who introduced him to Emilie de Châtelet and Voltaire. His quarrelsome nature may be indicated by the dispute that he maintained with Emilie (apparently about the non-payment of tutorial fees). Maupertuis apparently set aside this dispute and proposed König for membership of the Berlin Academy, but König repaid him in false coinage by attacking the Principle of Least Action and ascribing it, not to Maupertuis but to Leibnitz, stating he had evidence that this priority was claimed in a letter from Leibnitz to Hermann. When challenged, König stated he could produce a copy, but strangely the original was unobtainable. He said

Fig. 21 Pierre de Maupertuis painted by Tournières in 1743. He is dressed for the cold of Lapland. He rests his hand on the pole of the Earth, flattening it, and nearby (right foreground) there is a pair of dividers with which he will measure the scale of a degree. © Observatoire de Paris

that it was in the hands of a certain Swiss named Henzi who had been executed in Berne after being on trial for conspiracy; no trace of the letter could be found in Henzi's belongings. It was unwise to criticize the president of the Berlin Academy to the Academy on such unsubstantiated grounds and an internal hearing of the Academy reported negatively on all this. König resigned and withdrew back to his university in Bern.

Maupertuis was isolated and criticized by the Cartesians. He found approval, however, in the young mathematician Émilie, the Marquise de Châtelet, of whom he became tutor and lover. She worked with him on mathematics and on the explanations and commentaries which she wrote for her textbook on the *Principia* even after she had become also the lover of Voltaire (Bodanis 2006).

People Gabrielle Émilie Le Tonnelier de Breteuil, Marquise du Châtelet-Laumont (1706-1749)

Born in Paris into an aristocratic family, Émilie's father was a landowner and courtier. She was a bright child and her father educated her well. In appearance she was said by some to be large and ungainly and her big feet were especially noticed; as a grown woman, however, she was compared to a Greek statue in the gossip of the maid who had assisted her in her bath. As a child she was studious and became interested in mathematics and the sciences. In these disciplines she was encouraged by one of her father's many intellectual friends. She was described as of a "passionate nature," never lacking for romantic attachments, and at the age of nineteen she married Florent Claude Chastellet, a soldier and an aristocrat becoming the Marquise de Châtelet. The couple inherited the estate of Cirey, and Émilie de Châtelet had three children by the Marquis who was, however, frequently absent on military duty. The Marquise lived an extravagant social life in Paris, separately from her husband and surrounding herself with many men. It was both their intellectual stature and their love that mattered most to the Marquise; one lover was Maupertuis, who was her mathematics tutor. Émilie first met Voltaire when he was a house guest of her father. She met him again when he had just returned from England. Voltaire displaced Maupertuis, who however continued to teach Émilie mathematics and contentedly worked with Voltaire. Voltaire was eventually displaced in his turn by Clairaut, but the happy ménage was united in its opinion of the truth of Newton's world view.

At a later time in history Émilie may have become a member of the Academy of Sciences but as a woman this was not then possible. The less formal places for scientific discussions were the cafés of Paris, and although women were also not allowed in cafés, Émilie decided in 1734 that she wanted to go into the Café Gradot. This Paris café was most famous as a meeting place of mathematicians, astronomers and scientists, including Maupertuis. To enter she dressed as a man to parody the discriminatory rule, and she joined Maupertuis' table, who cheered and ordered her a cup of coffee. The proprietors pretended not to notice they were serving a woman – Émilie attended the scientific meetings at the Café Gradot regularly, fashionably dressed in man's clothes.

In 1734 Voltaire was forced to hide to escape a warrant for his arrest and went to live with Émilie at Cirey. Her husband was complacent at the arrangement, permitting Voltaire not only to sleep with his wife but also to spend money improving the house, building a gallery with a very fine collection of scientific instruments and a large library. With Maupertuis and Clairaut, and later with the Swiss mathematician, Samuel König, Émilie and Voltaire worked on books about Newton, philosophy and mathematics, both separately and collaborating together. Émilie translated the *Principia* into French, adding explanations to help the reader, some of which were based on discussions with her by Clairaut, and Voltaire wrote the preface. A dispute with König over his allegations that one of her books on Leibnitz was merely a rehash of what he had taught her ended their friendship and association.

In the spring of 1748, Émilie took another lover, Jean François de Saint-Lambert, a poet. While giving birth to his child in 1749, she died.

Maupertuis concluded that the debate about the shape of the Earth could only be convincingly resolved by measurements made nearer the equator and the pole and therefore overseas expeditions were necessary. In his campaigns in support of Newtonian theory, Maupertuis had the support of Voltaire, who portrayed him against the established French Academicians like Galileo against the Inquisition. This must have made Voltaire's betrayal of Maupertuis in later life (see text box) harder to bear.

The Academy had already accepted a suggestion by Charles-Marie La Condamine in 1733 (repeated a year later by Louis Godin) that it should organize a major expedition to the equator to measure the scale of the Earth there. In doing so the Academy had acknowledged that there were no accessible lands south of Paris near to the equator that were suitable for a surveying expedition because the Paris Meridian passes south into Africa through the Sahara desert and into the jungles of West African near Benin. In addition, it was impractical to think of an expedition to the East Indies since an expedition would have to penetrate into areas inhabited by war-like savages. The Academy therefore negotiated with Spain to send an expedition into Peru (the area now known as Ecuador) in South America.

When Maupertuis added that a second expedition north towards the pole was needed, it was even more impossible to make the measurements of the scale of the degree of latitude along the Paris Meridian since the Meridian passes from Dunkerque into the North Sea and east of England and between Scotland and Norway never touching land before it terminates at the North Pole. Visiting Paris in 1734, Anders Celsius (1701-1744), professor of astronomy at Uppsala in Sweden, whose name is remembered in the naming of the centigrade temperature scale, may have suggested that the expedition (1734-37) was sent north to the Arctic Circle to Sweden and into Lapland. A second expedition followed and was sent south towards the Equator to Peru (1735-1744).

THE NORTHERN EXPEDITION was led by Maupertuis into the Gulf of Bothnia, the northern extension of the Baltic Sea between Sweden and Finland. A portrait by Robert Tournières made on Maupertuis' return shows him dressed as a Laplander with fur hat, leaning on a globe at the poles and flattening it. The picture is labeled as follows: "It was his destiny to determine the shape of the world." In the engraving he is also pictured being pulled by a reindeer in a sled. The picture is fanciful and not as authentic as it looks, but it showed how Maupertuis wanted to present himself – as an adventurer, mathematician, explorer, and man of action.

Maupertuis was supported by Celsius who not only was an able mathematician, astronomer and essential part of the science team but also the liaison between the expedition and the Swedish authorities, on whose territory the work had to be done. The northern team also included the astronomers Alexis Clairaut and Pierre Le Monnier with assistance from Abbé Réginald Outhier, Charles-Etienne-Louis Camus and Anders Hellant (1717-1789), the expedition's interpreter who was brought up in Lapland. The expedition drew attention to what had been up to that point, and perhaps one could argue it still remains, one of the less visited parts of the world. The published journals of Maupertuis (1738) and Outhier (1744) made Lapland life and customs better known.

People Pierre Le Monnier (1715-1799)

Born into a distinguished family, Pierre Le Monnier was an astronomer and member of the Academy of Sciences from an early age. As a young man he worked on the theory of the orbit of the Sun, made an elaborate map of the Moon and observed the 1731 opposition of Saturn. In 1735 he joined Clairaut and Maupertuis on their expedition to Lapland to measure the meridian there. A youthful and vigorous man, he was able to contribute in large degree to the success of the scientific measurements. He not only participated in the surveying measurements and the triangulation process to establish the scale of a degree at that latitude in the north, but also worked to identify the effects of the atmosphere on the measurements. He returned to Paris and began a study of the orbit of the Moon, and he supervised the construction of the *meridiana* in the more civilized environment of the Church of St Sulpice (see Chapter 4). In addition to his many activities, Monnier constructed a transit instrument for the Paris Observatory and observed an eclipse of the Sun from Scotland.

When making a catalogue of the position of the stars prior to the discovery of the planet Uranus, he observed the planet 12 times, without recognizing its nature. In 1776 he added a new constellation to the sky, Turdus Solitarius, the "Solitary Thrush," but it fell from favor and was rationalized away, its stars absorbed into adjacent constellations.

The expedition members began their visit to Sweden in Stockholm where they were presented to the King of Sweden and gathered local knowledge from the Swedish office of mapping. In June 1736 the party split into two, one group of those more liable to sea-sickness traveling by land and the other by sea, to the estuary of the River Tornea (or Torne or Tornio or Tornionjoki), 1135 km from Stockholm. At the present day the River Tornea forms the natural boundary between Norway and Russia in the north and Sweden and Finland in the south (Regional Library of Lapland and the Pello Municipal Library 2006) and at the time of the expedition the area was all part of Sweden. They arrived in the summer, "time enough to see the Sun perform his course for several days together without setting, a sight that strikes with wonder an inhabitant of the Temperate Zone" (Maupertuis 1738).

The astronomers measured a meridian arc between the church bell-tower in the town of Tornio on a peninsula near the mouth of the river and the mountains north of Kittis. These places were linked by the river, which flows due south along a meridian arc. They traveled upstream, mounting the cataracts in the rough water in Lapp boats of shallow draft and made of thin, deal planks flexible enough not to be damaged by frequent collisions with submerged rocks but as a consequence very unstable.

They had planned before setting out from France to make measurements from islands along the river, but the islands proved to be very low lying and they could not establish lines of sight between them over large enough distances. Therefore they had to make forays, or "most painful marches," inshore from the river and climb mountains, "clambering up steep rocks" to get line of sight connections between the surveying sites. At each mountain top site they built a "target" which also served as a shelter for the surveying instrument. Each target was a hollow cone of pine trees tilted together to touch at the tops and from a distance tapering to a point. The trees were stripped of their bark to make them more visible, and the sighting telescope was mounted inside the cone on its axis to make the measurement outwards from the site. They used quadrants and sectors to measure the angles of the triangulation network that had been made by the instrument maker Claude Langlois (Fig. 22).

The geodetic survey took two months in the summer of 1736. Even then, the weather in the mountains was bitterly cold, much more so than in the valleys. In one instance they had to camp on a mountain for ten days and wait for the atmosphere to clear; the fog lifted only when the wind was from the north. They were also impeded by a forest fire that filled the air with a smoke haze and had to send someone to distant targets to cover them with a white sheet to see them better. One target was burnt down and they had to re-erect it. Fortunately they had had the foresight to mark the positions of the targets with stones and stakes. Maupertuis remarked on the "beautiful lakes" in the mountains that gave the countryside the "air of an enchanted island in a romance." He sourly added, "...anywhere but in Lapland it would be a most delightful place."

Fig. 22 Portable quadrant of 65 cm radius and a telescope, made by Claude Langlois about 1730, and similar to instruments used by Maupertuis in Lapland. © Observatoire de Paris

Out in the field, the astronomers endured the summer clouds of bloodsucking flies (gnats, midges) that were so insufferable the Lapps habitually left the inland regions and descended to the coast. The expedition members covered their faces with veils to keep the insects off, but they found that the gnats quickly flew in through the gaps when they passed bread under the veil to eat. The gnats swarmed over any food that was put on the table, so many that it was difficult not to eat some of the flies as well as the mutton. On the journey upstream they encountered native Lapps (the Samis), migrating with their reindeer herds which they used for all their needs and living in temporary camps of hide or rag tents. Maupertuis was not charitable toward the Lapps, describing them as ugly and their tents as wretched. Outhier was not much more charitable – he noted their habit of continuous begging and that "there is no harmony in their singing." But the expedition learnt from two Lapland girls to eat surrounded by dense clouds of insect-repeling smoke from a smouldering fire. In order to avoid bites while asleep, they "were obliged, notwithstanding the excessive heat, to wrap our heads in our Lappmudes (a sort of gown made of rein-deer skins) and to cover ourselves over with branches of firs, and even whole trees."

In October and November, the expedition measured the latitude difference between the two places at the extremities of their arc using an instrument called a *zenith sector*. Imagine a 90° frame consisting of a vertical post 12 feet (2.4 meters) long with an arm of the same length running horizontal from its top. The end of the horizontal arm and the foot of the vertical post are connected by a brass quarter circle on which a scale of degrees is engraved. There is an arm that carries the eyepiece end of a telescope along the scale: it is pivoted at the center of the quarter circle. The zenith sector is set up with the post vertical, aligned by a plumb bob, and the astronomer sights a star with the telescope to measure its altitude on the brass scale, tracks the star in its motion, watching the altitude rise and fall. The star is at its greatest height above the horizon when it is on the meridian line overhead and its altitude at that moment tells the latitude of the observing place.

The objective of the expedition was to measure the latitude difference between two places and the distance between them, and this depended crucially on the accuracy of the zenith sector. Maupertuis' instrument was made by George Graham, a leading instrument maker of London. Maupertuis especially admired the work of the English craftsmen and had sent Celsius to London to commission several of Graham's instruments for the expedition to Lapland. A calibration test of the zenith sector instrument by all five members of the expedition on a measured triangle (with a base 36 toises long at a distance of 380 toises) showed a measuring error of only 3.75 arc seconds; this was probably due to its thermal contraction at the low temperature (Chapman 1995). Nevertheless, the measurement of angles by the sector turned out not to be as accurate as the expedition thought.

People Réginald Outhier (1694 1774)

Outhier was the canon of the cathedral at Bayeux. He was an amateur scientist and a correspondent of the Cassinis, working with them on the geodesic survey of France in Normandy. He published an agreeable account of the Lapland expedition, noting the customs of the people, the geography, etc. (Outhier 1744)

As the winter set in so did the cold, torrential rain and snow; the temperature fell so low that it froze the liquid in an alcohol thermometer. The result of the Lapland expedition owes much to the persistence of the astronomers using the instruments. It cannot have been pleasant, for example, to apply an eye to a frozen brass eyepiece or to clamp brass screws with bare fingers in the Arctic temperatures. I can speak of the pain from personal experience, having left a ring of skin from around my right eye frozen to an eyepiece when I withdrew my head from a telescope in an upstate New York observatory in a night-time winter temperature of 30 degrees Fahrenheit of frost (I put salve on the wound, closed down the telescope and went to bed even though it was clear and the stars were bright). Le Monnier had a similar experience in Lapland when his tongue froze to a silver cup from which he was drinking brandy at -20°C.

People Charles-Etienne-Louis Camus (1699-1768)

A little-known scientist, Camus was a clockmaker, and later an administrator at the Academy of Sciences. He was the scientist of least stature in the expedition to Lapland.

The surveying base was laid out on the winter ice and by December the river had frozen hard enough to bear the scientists' weight. Like Picard's choice for the location of his baseline on the plain where Orly airport is now, the surface ice was mostly level, but there were obstructive blocks of ice and drifts of snow over its length of 7400 toises (14 km). These had to be cleared so that the standard measuring sticks could be laid down from end to end, and Maupertuis, Le Monnier and the others were only partly successful in clearing the baseline. They pulled a triangular snow plow made of large logs and dragged pointy end first to scrape the snow aside but it did not sink deep enough into the snow. In another attempt they harnessed yoked cattle to the two ends of a tree trunk but the log easily bounced over obstructions; in the end the baseline had to be cleared by hand. They worked by twilight and the light of the aurora reflecting on the snow, watched by Laplanders drawn by the novelty of the sight. As a standard measure, they had taken an iron toise that had been calibrated against the standard at Paris (it became known as the Toise du Nord), and they used this to make five-toise measuring rods of fir. The rods were heavy and had to be laid out carefully on the snow. The scientists did not trust their helpers to butt the ends of the rods together carefully enough and always did these tasks themselves, noting each rod down on papers clipped to boards strung around their necks. Like Picard, the expedition members guarded against the possibility that, in the dark and the cold, they had miscounted the successive placement of the rods. They did this by checking the length of the standard baseline with a cord that they had measured but that stretched. This was not as accurate as the rods but was longer and there was less chance of making a counting mistake.

As the winter drew in, it became practically impossible to work outside since the temperature dropped to 15 to 25 degrees of frost. Maupertuis (1738) described the icy labor:

> Judge what it must be like to walk in snow two feet deep, with heavy poles in our hands, which we must be continually laying upon the snow and lifting again, in cold so extreme that whenever we would take a little brandy, the only thing that could be kept liquid, our tongues and lips frozen to the cup and came away bloody, in a cold that congealed the fingers of some of us… While the extremities of our bodies were thus freezing, the rest, through excessive toil, was bathed in sweat. Brandy did not quench our thirst; we must have recourse to deep wells dug through the ice, which were shut as soon as opened.

During the summer's work, the expedition had forgotten to take a crucial measurement (the height of something that they had used to support instruments), and Maupertuis returned to the mountain top in the snow to rectify the situation. He remarked on the local methods of travel over snow that he had to adopt with a turn of phrase that brings to mind images of the scientist tumbling head over heels; it was the first time that he had used skis or a reindeer sledge. In order to walk, or rather slide along, he wrote that the Lapps and Finlanders use "two straight boards eight feet in length … to keep from sinking into the snow. But this way of walking requires long practice." Maupertuis' boat-shaped sledge was pulled by a half-wild reindeer (Fig. 23) but Hellant, the interpreter, was practiced; he balanced well in his sledge while Maupertuis drove at high speeds and often tumbled out. This infuriated the reindeer but Maupertuis found that he could turn the sledge upside down

Fig. 23 Maupertuis used a reindeer sledge to travel over the snow. It was shaped like a boat and he used a stick to stop it toppling over as the reindeer bounded along. He also traveled by skis like the person on the side of the mountain in the distance. Outhier (1744)

to shelter himself from its kicks. Outhier (1744) says, "Some travelers have pretended that, on being told in its ear the place to which you were disposed to go, the reindeer understood you; this is a mere tale."

When Maupertuis returned to Tornio, the town showed "a most frightful aspect" with its little houses buried to the tops in snow, so that if there had been any daylight it would have been completely shut out. It was so cold that if the scientists opened the door of a warm room to the outside, the cold air rushed in and caused it to snow from the humid air inside the room. The expedition members rested up at Tornio until March 1737 before continuing on.

Tornio was composed of three streets with about 70 wooden houses. The town was small, but the people were clean, persistent, skilful, educated, and hospitable. The young men in the expedition got on well together, and the party had an atmosphere of youthful exuberance, the subject of gossip back in Paris but welcomed in Tornio. The Frenchmen were unimpressed by the food: dried fish, oatcakes, salmon, mutton seasoned with sugar, saffron, ginger and lemon, and orange-peel. They sometimes declined to eat what they were given. They drank water, sour milk, ale, and occasionally local wines; white wines were common but red wines were scarcely known, and the locals thought that the Frenchmen were drinking "sheep's blood." The saunas, with their accompanying rituals of whipping (while naked) with twigs and walking in the snow, were a source of breathless amazement. Maupertuis tried a sauna at a temperature scientifically recorded of 44° Réamur (about 55°C).

The Abbé Outhier records the religious orthodoxy of the townspeople with approval, noting that it was the duty of a verger in the church to prod slumbering members of the congregation awake with a stick and that church services at Easter lasted practically all day. Officers had the right to enter houses to see that the chimnies were properly maintained and that a lantern was kept ready. The Frenchmen, as Roman Catholics, were not permitted to hold religious services in public, this being forbidden by the strong Lutheran law. The Frenchmen had brought an interpreter but were able to converse directly with the mayor, the governor, the rector, and the local school-teacher in Latin. It was the teacher who had suggested the surveying route and triangulation points, and by the end of their stay some Frenchmen had learned enough Finnish to get by.

Two residents of Tornio were particularly affected by the expedition- the Planström daughters, Elizabeth and Christine. When Maupertuis and his colleagues returned to France, the two women traveled with them, survived the journey's many dangers, and became Roman Catholics. Christine eventually entered a convent, disturbed, one can surmise, because after a passionate *affaire* Maupertuis lost interest in her; a love poem from Maupertuis to her survives, written, presumably, on the spot in Lapland during Arctic long summer days, cold weather and the aurora. Its literary merit is suggested by the Lesley Murdin translation given here in the overblown, mawkish spirit of the original. Christine's sister, Elizabeth, married a Frenchman, M. de Pelletot. There was talk that perhaps the main motivation of her husband in marrying her was associated with her substantial dowry, and, indeed, in 1761 the marriage ended in an unhappy divorce. She had to fight a hard battle for the right to acquire alimony (3,000 livres per year) to support her and her one son, and in the end, she too took to a convent at Rouen.

Ideas Poem for Christine by Pierre Maupertuis, translation by Lesley Murdin

Dans les frimas	Here we must go
De ces climats	Where the icy winds blow
Christine nous enchante:	Christine can entrance us:
Oui, tous le lieux	Yes, the sun can rise
Où sont tes yeux	In the fire of your eyes
Sont la zone brûlante.	Where warm light dances.
L'astre du jour	Staying under this star
A ce séjour	The gloom stretches far
Refuse la lumière;	And blots out the light;
Et tes attraits	But all the action
Sont désormais	Of your attraction
L'astre qui nous éclaire.	Makes our star radiant bright.
Le soleil luit:	Summer brings light
Des jours sans nuit,	Days without night,
Bientôt il nous destine;	With that we are serene;
Mais ces longs jours	But these long days
Seront trop courts	To sing her praise
Passés près de Christine.	Are too short for Christine.

During the winter in Torino, the expedition members played hard and, it seems, loved well, but they also worked. For example, they made observations of a pendulum and found that the pendulum they had brought from Paris accelerated 59 seconds per day at that latitude. They reduced their data and found, to their astonishment, that the length of a degree at this location, on the Arctic Circle, was getting on for about 1000 toises more than the Cassinis had calculated by extrapolating their results over the latitude difference of France. This was so unexpectedly large that Maupertuis felt some of the observations had to be verified in the spring. The geodetic surveys had been observed many times by several independent persons and Maupertuis had caused measurements to be made of all three angles of some triangles. The three angles have to add to 180°, and they did, so Maupertuis was confident of the triangulation. The party re-measured the star positions and the latitude of the two extremities, and there was apparently no mistake – the length of a degree of latitude at the Arctic Circle was 57,438 toises, 377 toises longer than Picard had measured. As it later transpired, they had reason to doubt the size of the proof that the Earth was flat at the poles.

As would be natural within a scientific expedition, the party took great interest in the curiosities of their up-to-then largely unexplored environment. Maupertuis described the aurora:

> When I look into the sky there is a fabulous spectacle. Fires of a thousand different colors light it up, making ripples like drapes across the sky... It is from such explorations that we will understand the Universe.

As they prepared to depart Torino in April 1737, the party took time out to visit the Käymäjärvi Inscriptions, a stone engraved with an undeciphered runic inscription, located near Lake Käymäjärvi. Maupertuis later gave an account of the stone to the Academy. The expedition left Lapland early in June as the ice broke up leaving the sea clear to sail.

The expedition to Lapland was not simply uncomfortable it was extremely dangerous. Le Monnier fell ill during the winter and Maupertuis' health was permanently damaged. The expedition was shipwrecked in the Baltic Ocean on the return journey. The wind got up soon after they had weighed anchor and the ship took in water. They tried to bail out and pump it dry by throwing the cargo of wood overboard to get more clearance above the surface of the sea. These efforts were not successful and after three futile days of this labor, the pilot carefully ran the boat aground. The party salvaged the papers and the instruments (although they were damaged by immersion in sea-water) but no lives were lost. It took a week to repair the boat, but they continued their journey on to Denmark and through the Low Countries back to France.

MAUPERTUIS RETURNED to Paris in August 1737. He had sent nothing of his results ahead, and the Academy was eager to hear them, convening a meeting on the 28th of August for this purpose. It caused a sensation, and the news that the Earth was flattened as Newton had calculated spread immediately from the Academy into Paris, to the rest of France and to England. The Cartesians were in consternation at the information and the Newtonians jubilant. Mairan indignantly wrote that it was the astronomers of the Academy and of France that had taken the lead in establishing the shape of the Earth and "there is no justification, it seems to me, for the English to be making noise about this."

Cassini and his entourage regrouped to attack both Maupertuis' result and Maupertuis himself. In the Academy, Cassini pointed to some weaknesses in Maupertuis' techniques; he had measured quite a short distance, not even a full degree of latitude; he had measured only the fundamental baseline at the southern end of the triangulation chain, and not measured a baseline to check the northern end; the astronomical measurements had not been re-calibrated at their finish.

Cassini also attacked Maupertuis personally. He inspired, perhaps encouraged and possibly even helped write an anonymous pamphlet (*Physical and Moral Stories*). The pamphlet stoked the fires lit by Cassini in the Academy and moved rapidly from the technical deficiencies of the measurements to allegations that the work exposed France to English ridicule. In a musical *double entendre,* the pamplet accused Maupertuis of debauchery in Lapland: "He thought only of finding some pretty girl to pass the night playing her guitar." Even more so, on the authority of the Abbé Outhier, the pamphlet accused Maupertuis of being a bad example to the whole expedition: "Maupertuis' colleagues followed the example of their leader, and each took a mistress."

The Planström sisters, Elizabeth and Christine, were visible and relevant evidence for the allegations of misconduct, even if the attacks were malicious, exaggerated and had nothing to do with the science. Most of Maupertuis' colleagues kept their counsel and of them only Celsius attacked back. Voltaire responded in his characteristically vivacious style, in the book *Elements of Newtonian Philosophy* (1738).

The message that the Lapland expedition brought home was clear: the length of a degree in Lapland was longer than the length of a degree in France. « *Le degré du Méridien qui coupe le Cercler Polaire surpassant le degré dur Méridien en France, la terre est une sphéroïde applati vers les Poles*, » Maupertuis wrote : "The degree on the meridian that cuts the Arctic Circle being larger than the degree on the meridian in France, the Earth is a spheroid flattened at the poles." The result was acclaimed and Maupertuis triumphed over his scientific opponents. Voltaire referred to Maupertuis as the "flattener of the Earth and the Cassinis."

In reality, though, this result was too decisive because Maupertuis' measurement (that the degree in Lapland was 500 meters longer than the degree in France) was too much. "This flatness [of the Earth] appears even more considerable than Sir Isaac Newton thought it," wrote Maupertuis. Johann Bernouilli, on the other hand, was sceptical, thinking that Maupertuis was biased: "Do the observers have some predilection for one or other of these ideas? Because if they believe the Earth is flattened at the poles they will surely find it so flattened. … [Therefore] I shall await steadfastly the results of the American observation." The Scottish mathematician James Stirling said that he too would remain neutral about the result, "till the French arrive from the South, which I hear will be very soon." He overestimated the speed, though, with which La Condamine would return from South America.

Maupertuis' measurement was repeated in 1801-1803 by the Swedish geodesist Jöns Svanberg, and the resulting length of a degree in Lapland was considerably smaller than Maupertuis had obtained. Svanberg determined that the degree at Tornio was 57,196 toises long, 400 meters shorter than Maupertuis determined so the Earth was flattened but not by as much as Maupertuis originally thought. According to the Finnish scientist Yrjö Leinberg in 1928, Maupertuis' too-large a result was due to the following principal factors: there was an error in the triangulation the equivalent of 45 meters in the distance between Tornio and Kittis, and there was an error due to the "deflection of the vertical." This was caused by the area's geology- extra dense rocks on one side of the plumb line or another deflect the plumb bob causing the instruments to be misaligned to the vertical. The largest error was in the sector instrument which was altered during transportation. There were six sailors assigned to carry the sector on the expedition up-river, and considering the instrument's size, weight and the cataracts and rocks over which it was manhandled, it would not be surprising if it had taken a knock or two.

People Charles Marie de La Condamine (1701–1774)

Trained as a mathematician, La Condamine (Fig. 24) was at first a soldier but found this life did not suit him and therefore took up academic work in Paris. This did not suit him either, and he embarked on a journey of exploration in the Levant, documenting his discoveries upon his return in a volume that impressed the Academy and led to his selection to lead the Peru expedition.

On his return to Sweden, Celsius also found fame and a tangible life pension of 1000 livres per year. He used his experience in Lapland to develop a thermometer, which he marked with a 0 (at the boiling point of water) and 100 (at freezing). The reversal of the scale to 0° representing freezing point and 100° representing boiling point was made by Linné, but it was credited to Celsius whose name is now used for the metric unit of temperature.

Maupertuis' expedition to Lapland was commemorated by a Finnish postage stamp in 1987. There are memorials to him and the expedition near the church in the town of Tornio, now twinned with Haparanda in Sweden, and together known as *Eurocity*.

People Pierre Bouguer (1698–1758)

A child prodigy, Bouguer succeeded his father as a professor of hydrography at the age of 15. He quickly distinguished himself at the Academy and accompanied La Condamine on the trip to Peru to measure the shape of the Earth. Apart from outstanding work on the expedition, he investigated a wide variety of other scientific phenomena. His interest in the measurement of light dates from 1721 when he answered a question posed by Mairan on the relative amount of light from the Sun at different altitudes. He constructed a photometer that used a candle as a source with which to compare the object under investigation. The candle was attenuated by altering its distance from the surface that it was illuminating until its brightness looked the same as the target source and then using the inverse square law to measure the dilution of the candle's light. He used the eye as a detector only to make the comparison and consequently invented the null photometer. He also established the exponential drop-off of the intensity of a beam of light as it passes through a uniform medium, e.g. air or water. This is known as Lambert's law, since Lambert restated it in a book that was more widely circulated.

WHILE MAUPERTUIS was suffering from the terrible cold of an Arctic winter, the Peruvian expedition was sweltering in equatorial South America (Whitaker 2004). The principal French scientists Charles-Marie de La Condamine, Pierre Bouguer and Louis Godin.

Fig. 24 Charles-Marie de la Condamine, painted in 1761. © Observatoire de Paris

People Louis Godin (1704-1760)

An astronomer, Godin (Fig. 25) worked on many workaday matters for the Academy of Sciences, such as the preparation of the *Connaissance de Temps*, and made scientific studies of meteors, eclipses and pendulums, as well as working with the Peruvian expedition to measure the shape of the Earth. He was of high seniority but of the least scientific stature in the Peruvian expedition and was estranged from Bouguer. He remained in Peru after the expedition had left and was a teacher of mathematics at the University of San Marcos. After a brief return to Paris he went to Spain to teach naval cadets and died in Cadiz.

The French party was accompanied by two Spanish naval officers, Jorge Juan and Antonio de Ulloa. Their participation was a result of negotiations in 1734 between King Felipe V of Spain and his cousin King Louis XV of France. The equator, at which the length of a degree was to be measured, passes across the northern part of South America through what is now Ecuador (the name of the country is Spanish for "equator"). In the eighteenth century the area was part of the province (*audiencia*) of Quito, part of the Viceroyalty of Peru, and directly ruled by Spain therefore an official French expedition into Peru could not take place without the permission of Spain.

People Jorge Juan y Santacilia (1713–1773)

Born into a noble family and educated in Malta, Juan rose to the rank of captain in the Spanish navy. He was selected for the expedition to Peru on the basis of his scientific interests. He made a great success of his participation in it and was honored by being elected to many scientific societies. He was then given a variety of scientific assignments for the Spanish navy, including an espionage expedition to England to learn the details of English ship construction. He became Spanish Ambassador in Morocco.

Fig. 25 Louis Godin painted by Jeaurat. He caresses a terrestrial globe at the equator of South America, sensing its shape with his finger tips © Observatoire de Paris

People Antonio de Ulloa y de la Torre Giral (1716–1795)

Ulloa became a sailor at the age of 14 and eventually rose to the rank of admiral in the Spanish navy. He volunteered for the Peru expedition after someone else dropped out and an opportunity opened. While on the expedition, he was recalled to naval duty for four years to fight against England and rejoined the expedition as it ended, with only Godin left to complete some of its non-geodetic aims. Separate from the expedition, he discovered the metal platinum by separating it from gold and silver. On his return from the Peruvian expedition in 1745 he was captured by the British navy and was imprisoned, first in Louisburg, Canada, and then in England. He seems to have been at liberty within London and was made an associate of the Royal Society, which helped him return to Spain where he was assigned various duties for the Spanish Navy. He was also governor of Spanish Louisiana but was expeled by the French colonists, then assigned to a voyage to the Azores with scaled orders to proceed to Havana and lead a campaign intended to recapture Florida for Spain. Preoccupied with scientific observations, he forgot to open his sealed orders and returned to Cadiz after a cruise of two months. Presumably surprised to see him back so soon, the authorities arrested and tried him. He was acquitted but assigned to land duty in charge of a military academy. Ulloa established the observatory at Cadiz, and his name is remembered as a meteorological phenomenon called Ulloa's Halo, a ring of light around the antisolar point seen in mountain mist.

King Louis XV wrote to his "dear uncle" on April 6, 1734 that there was no reason for the Spanish to be suspicious of French motives because his mapmakers would be making observations "which would be advantageous not only for the advancing of science, but also very useful for commerce, by increasing the safety and ease of navigation." The observations were mainly for the sake of pure science, he claimed, and would have little practical implications for navigation but the French king was being disingenuous. A brief about the proposed expedition to King Louis from his Minister of Marine, Count Jean-Frédéric Phélypeaux de Maurepas, clearly indicated the intelligence opportunities: a scientific expedition would be above suspicion and would enable France to study the country and bring back a detailed description.

Felipe was advised by those in Madrid still suspicious about the possibilities that the real motives of the French expedition into the Spanish territory were imperial. In his agreement, Felipe set conditions that the expedition should present its equipment and stores for inspection upon entry into the Viceroyalty and at significant stops, it should not enter into illegal commerce and it should follow the prescribed route. He ordered two of his officers to accompany the expedition and keep an eye on what was happening; what in France some thought to be a humiliating condition, was in Spain seen as a wise precaution.

However, Felipe gave every indication that he had signed up for the scientific motives of the expedition. He wanted the Spanish officers to be full participants and offered to pay half of the cost of the expedition. He also made monetary drawing rights available in Peru to provide operating funds, keeping a measure of control over the expedition and rights to the benefits from it. He was genuinely helpful in offering negotiations for purchases in Peru in order to avoid exploitation by the strangers, but whatever the truth of the politics, the Spaniard officers were fully motivated and made a great scientific success with their participation.

The French party left Rochelle in May 1735 and arrived at Carthagena in Colombia in November to pick up the Spanish officers, reaching the Atlantic coast of Panama before Christmas. They crossed the Panama isthmus by boat and by foot continued to Panama City then set sail south in the Pacific Ocean. By the end of March 1736 they had arrived at the Gulf of Guayaquil, but unlike the Lapland expedition, which worked as a single team, the scientists in Peru split into separate expeditions (Godin with Juan, and La Condamine and Bouguer with Ulloa). This split reflected the uncertain management arrangements in the expedition; Godin was a senior member of the Academy and through seniority was appointed the nominal leader of the expedition. Bouger was the most distinguished member of the expedition (when considered as a scientist) and was next in command but gave no loyalty to his leader; even before the expedition had left Paris, Bouguer had complained to the Academy about Godin. Bouguer thought Godin incompetent and the weakest of the expedition's French scientists. La Condamine had first suggested the expedition and had taken all the initiatives to see that it progressed, but he was a junior member of the Academy and to both Godin and Bouguer. Nevertheless, he carried out most of the organization of the expedition.

Bouguer's relationship with La Condamine was very strained and the two men had completely different personalities. Bouguer was interested only in science and was happiest with his head in mathematical calculations. In contrast, La Condamine was a practical man, curious, and even an entrepreneur; he opened a trading post in Quito for silk and lace when it turned out that the drawing rights he could access provided limited money. Godin progressively withdrew from both Bouguer and La Condamine and associated only with the Spaniards, working in secret. Allied with neither side, the Spaniards were able to keep the expedition together. When the expedition drew to a close, though, Godin was to stay behind in Lima to teach. When the Academy, examining the chaotic finances of the expedition, came to suspect embezzlement, Godin's decision not to return to France looked suspicious.

At that time, the city of Quito was the inland capital of the province of Quito, now the capital city of Ecuador. On the Yarouqui plain near Quito, the expedition laid out the fundamental baseline of 6,300 toises by hacking away the scrub in a straight line between two end markers. They measured the baseline with three-toise standard wooden measuring rods, made of wood tipped with copper so they could be butted accurately together. One group led by Bouguer, La Condamine and Ulloa, measured in one direction and the other led by Godin and Juan measured in the opposite direction; their measurements differed by only 8 centimeters. They then surveyed Ecuador along its north-south running mountain range extending over 3° of latitude and more than 300 km from Quito in the north to Tarqui in the south. They laid out a check baseline of 6196.3 toises near Tarqui, part of it measured across a shallow pool with the standard rods floating on the surface. They were gratified to find that it disagreed with their triangulation of the same distance by only 0.2 toise (40 cm).

Out in the field making their measurements, triangulating from mountain to mountain, the scientists braved hostile and dishonest natives, snakes, scorpions, and mosquitoes, in addition to freezing deserts and active volcanoes (Bouguer complained at one station that "sleep was continually interrupted by the roaring of the volcano," a "frightful noise"). Weather conditions in the mountains were particularly difficult because the peaks were often shrouded in mist, making it impossible to see from one mountain triangulation station to the others. Overnight the scientists were sometimes frozen into their huts by snowfall blocking the doors and had to rely on visits by their native assistants to break them free from the outside. Even worse were storms so violent that the survival of the surveying parties was often in doubt. After a particularly violent storm, for example, public prayers were offered for the safe return of one group, "or at least for someone to re-assure us," said La Condamine. Ulloa (1748) offered insight to the physical dangers of the mountain expedition:

> Our common position was inside a hut because the extreme cold and violent winds did not allow anything else. We were continuously enveloped in such a dense cloud that there was nothing to see … When the clouds engulfed us, our breathing was made difficult by a greater density of the air, the continuous fall of thick snowflakes or hail, violent winds and a continuous fear that either our living quarters would be uprooted, throwing it and us over the nearby precipice, or that the weight of ice and snow which accumulated quickly on the hut would cave in and bury us … We were frightened by the rocks which came crashing down when they became loose. In their fall they not only caused the entire peak to shudder, but they swept along everything that was in their path.

Some of the expedition's mountain-top measurements were the highest scientific measurements made up to that time. On one mountain, Rucu Pichincha (15,413 feet high, 4.7 km), they were shocked to see that the barometer stood 300 mm below the normal height of 760 mm of mercury, the lowest atmospheric pressure that anyone had ever observed. It was little wonder then that on the ascent they suffered vomiting and fainting, what we know today as altitude sickness, due to lack of oxygen in the rarefied air. This mountain defeated their geodesic measurements and after 23 days they retreated to a lower altitude to complete their telescopic observations of other distant peaks. The highest altitude at which they worked was on Mount Corazon at 15,794 feet (4.8 km), whose altitude they measured to within a few toises (say 10 meters). This is the same height as Mont Blanc, first climbed in August 1786, 50 years later. On Mount Corazon their "clothes, eyebrows and beards were covered with icicles," and it was little wonder that the members of the expedition were amused when, a few months later, they received a letter from their colleagues in France expressing concern that, so close to the equator, they must be "suffering too much from the heat."

Each expedition observed natural phenomena, as well as the geodetic measurements that were the initial objective of their trip such as optical phenomena in mists and fogs, like haloes around the Sun (describing for the first time the *Ulloa Halo* or *Bouguer Halo*), Inca ruins, volcanoes, trees, animals, and birds. Bouguer in particular made observations of the period of a pendulum in the mountains, and he discovered for the first time the influence of the density of local rocks on the Earth's gravity; the difference between gravity at a given place and the mean is still called the *Bouguer anomaly*.

The lives of the expedition members were not completely uncomfortable. In Riobamba, for example, they were royally entertained as distinguished visitors from Europe, especially by the Peruvian women who were keen to learn the latest dance steps from France. La Condamine was especially charmed by the musical abilities of the oldest daughter of the family of Don Joseph Davalos. However, he was shy because he had facial scars from smallpox that he felt were disfiguring, and she had the sole ambition to become a nun. Needless to say, their relationship did not develop.

In Tarqui they were amused by a native festival; the members had observed the astronomers crouching at the eyepiece of a telescope observing the Sun to determine time to measure the latitude and longitude of one end of their meridian line. La Condamine wrote:

> It must have been for them an impenetrable mystery, to see a man on his knees at the base of a quadrant, head facing upwards in an uncomfortable position, holding a lens in one hand and with the other turning a screw at the foot of the instrument, and alternately carrying the lens to his eye and to the divisions to examine the plumb line and from time to time running to check the minutes and seconds on the pendulum and jotting some numbers down on a piece of paper and once again resuming the first position. None of our movements had escaped the observations of our spectators and when we least expected it they produced on stage large quadrants made of painted paper and cardboard which were quite good copies, and we watched as each of us was mimicked mercilessly. This was done in such an amusing manner that I must admit to not having seen anything quite so pleasant during the years of our trip.

There were also delicate non-scientific matters to deal with. La Condamine was instructed by the Academy to leave an inscribed monument commemorating the expedition, and near Quito he erected two pyramids, one at each end of the fundamental base, to mark it permanently. They were sizeable monuments, 4 meters square rising to 5 meters high. The text of the inscription had been drafted by the Academy before La Condamine left France and caused internal difficulties within the party. The inscription identified the principal French participants but not the Spanish officers in enough prominence, referring to their role (in the several versions) as "assistants" or "with the cooperation of…" It also did not mention the Spanish institutions equally with the French and the Spanish officers were vociferously offended. Furthermore, the inscriptions on the pyramids were crowned by a French royal symbol, the *fleur de lys*. This raised questions of sovereignty- were the monuments a subtle claim of territory, or could they be used as such at some future point? The incident escalated into one of national pride and injury "to the Spanish nation and personally to the Catholic King." Locals dubbed it the "war of the pyramids" and sided with the French in order to humiliate the Spanish (since Ulloa and Juan were representatives of the disliked colonial power). In due course, however, and after three legal hearings, there was agreement within Quito about what was inscribed on the pyramids but they were later destroyed by orders from Spain. In 1836 though they were restored (not in the right places- their positions have been lost- but at the Observatory in the capital of Ecuador).

The Spanish officers were repeatedly challenged by the colonists and Creole inhabitants of Peru. They were derided as *caballeros del punto fijo* (gentlemen of the fine point). This described them literally as gentlemen who had swords, but also derided them as legalistic representatives of the colonial power. It was also a *double entendre* undoubtedly intended and understood to be an insult to the Spanish officers' manhood. In another episode, the Spanish officers were accused of illicit trade in Quito and avoidance of duty on imported items; it took La Condamine's skills in diplomacy to solve this difficulty. Later the Spaniards were ordered from Spain to interrupt their scientific work and walk the long road from Quito to Guayaquil to help prepare the coasts and towns against the attacks of the English navy and to build and command two frigates.

A SERIOUS INCIDENT occurred in Cuenca. The expedition doctor, Dr Jean Senièrgues, had been busy both treating the townspeople, as was his custom, and working at making a fortune in the New World. He treated a man in Cuenca, Francisco Quesada, whose daughter, Manuela, had recently been jilted by her fiancé, Diego de Leon. A condition for the separation was that her family should be compensated by a monetary payment. Senièrgues was asked to negotiate the payment but unwisely moved into Quesada's house, calling into question whether he was personally interested in Manuela. His morals were denounced from the pulpit by a priest, Juan Jiménez Crespo.

Diego de Leon had local allies and when the doctor attended a public festival with Manuela on his arm, a murmur of disbelief and disapproval arose from the crowd of 3,000 people. He was reprimanded by the festival's director, who told him his provocative behavior was disturbing the festival. The doctor retaliated by questioning the director's authority and by threatening the man with a beating. The director declared that the festival was canceled: "There was nothing," wrote La Condamine, "that could infuriate the common man more." With cries of, "death to the government and death to the French," an armed mob surged forward towards the doctor. Bouguer, La Condamine and the others tried to defend the doctor but were outnumbered and intimidated. They withdrew under a shower of stones with minor injuries (Bouguer was wounded in the back). Stabbed with a dagger by the festival's director, Senièrgues died of his wounds four days later. Two years after the incident, La Condamine succeeded in having the perpetrators tried and convicted and then was outraged when they were let off with a light punishment.

If the expedition had run into difficulties in Quito and Cuenca, the astronomical observations at the southern station of Tarqui were not free of difficulties either. To observe the elevations of the stars in their motions and thus the latitude of the southern station, a 12 foot zenith sector had been brought from France, but the instrument lacked rigidity. It had been separated into two pieces for ease of transportation, but the screws had been lost and it was fastened with strengthening bars and wires. The performance of its telescope also left much to be desired; its lens suffered from severe chromatic aberration, and it showed stars as colored halos without a distinct focus. Godin and Juan therefore made their own instrument with which to observe at Cuenca- in fact they made two, abandoning the first as unsatisfactory.

Given the bickering between those in the expedition, the difficulties of the mountainous terrain and in communicating with France (which was half a year's voyage across the Atlantic Ocean), the meager finances, and diplomatic difficulties it is perhaps not surprising that the expedition took ten years and stretched the patience of the French government's support. It must have been disheartening to be told in a letter from Maurepas (received in 1738) that Maupertuis had solved the problem of the shape of the Earth with the Lapland expedition. What now was the point of the expedition to Peru? The letter containing the news hinted that the expedition should return without giving explicit orders.

The letter was uncertain in its intent because no one in the Academy could bear the loss of face if the expedition was abandoned and after all the effort returned empty-handed. The members of the expedition were very disheartened by a phrase in another letter from Clairaut. It was meant to be encouraging. It said that the measurements in Peru were "vital to confirm Maupertuis' measurements." What had started as an expedition to solve a problem had been downgraded to an expedition to confirm someone else's solution. The expedition stayed to complete the survey, but La Condamine broadened the scope of the exploration to give more emphasis to natural history and other natural phenomena.

TEN YEARS AFTER they had left Paris, the principal French members of the expedition returned, with Jorge Juan. La Condamine arrived eight months after, via a dangerous trip through the Amazon basin, of which he published an account in 1745 entitled *Relation abrégée d'un voyage fait dans l'intérieur de l'Amérique Méridionale* (*Brief account of a voyage made in the interior of South America*). He collected natural history specimens that he gave to the naturalist Georges Buffon (1707–1788) for his works on natural history entitled *Discours sur la manière d'étudier et de traiter l'histoire naturelle, Théorie de la terre* and *Histoire des animaux* (*Discourse on the Way to Study Natural History, Theory of the Earth* and *History of Animals*) all three of which were published in 1749.

Among his contributions to natural history based on his South American expedition, La Condamine had already rediscovered rubber in the jungle near Quito and used it to make a waterproof case for his quadrant. His discovery with the greatest impact was quinine. It had long been known in Europe from native Inca usage that the bark of certain trees was effective against fever. It was sold under the name of the "countess's powder," after its use in 1638 to cure the Countess of El Cinchon, Francisca Henriquez de Riviera, and wife of the viceroy of Peru. It was also known as "Jesuits' bark" because she charged the Jesuits with its distribution in Peru as a medicine and as "Cardinal's powder" since it was shipped in large quantities to Europe by the Jesuit Cardinal Juan de Lugo, but there were several trees of the same or similar names and some of the powders were useless. According to La Condamine, the Incas had discovered the medicinal properties of the tree after an earthquake tumbled numerous quinaquina trees into a lake, whose waters became healthy to drink, and he identified the tree called by that name (its scientific name being *Cinchona officinalis*). Its bark had three colors and the red bark was both the most bitter and more effective against malaria.

Bouguer arrived in Paris in June 1744 but Godin and Ulloa remained behind in South America. Godin had married a Peruvian girl, Isabel Grameson, but after La Condamine and Bouguer had left, Godin decided to return to France. He traveled along the Amazon in 1745 to French Guyana to make preparations for the voyage to Europe and there became separated from his wife for twenty years when the borders were closed by political difficulties. Eventually Isabel made a heroic journey down the Amazon to rejoin her husband in 1770 and the two of them set foot in France only in 1773. Isabel's story was a sensation. Her journey would have been amazing if she had been a man and was an almost incredible adventure by a woman of the time. (Whitaker 2004)

Meanwhile in Paris a disagreement had arisen between Bouguer and La Condamine whose relations had been progressively deteriorating throughout the expedition. Apparently this sprang from an account by Bouguer of La Condamine's intention to measure a degree of longitude rather than of latitude until he had been told more explicitly what to do in orders from France. Bourguer implied that La Condamine didn't understand the purpose of the expedition at all, and that in fact, La Condamine had thought that by measuring a degree of *longitude* he could determine the shape of the Earth because, if the Earth was flattened, its circumference through the poles would be larger than its circumference around the equator, and the reverse. In the end it was decided that the difference was too slight to be measured critically enough and the measurement of a degree of longitude was canceled.

The argument between Bouguer and La Condamine became a bickering quarrel over trivial matters such as who had suggested what improvement to techniques and equipment at what time: "Bouguer could not disguise his feelings of superiority as a mathematician over La Condamine," wrote Jacques Delille in La Condamine's obituary, "He felt he should be the primary object of public affection." La Condamine seems to have bitterly regretted the loss not only of the ten years in preparing for and making the expedition but also the ten years of vexation afterwards.

Although by the time the explorers had returned to France in 1744 the issue of the shape of the Earth was regarded as settled, at least their results were regarded as confirmation. Unusually for a campaign organized by the Academy, the principals all published individual accounts, presumably because of the breakdown in their relationships and because of the national differences; it was necessary for the Spanish scientists, for example, to publish their own independent account to show how their contribution was of equal status to that of the French. Ulloa's and Juan's account "proved influential in an unanticipated respect: despite the austerely scientific nature of their enquiries into geography, geology, hydrography, and climate, they illustrated their work with stunningly beautiful diagrams that incorporated romantic representations of American landscapes: the first stirrings of what was to be a long story- the nourishing of European romanticism with images of America" (Fernández-Armesto 2006).

Bouguer's value for the length of the degree of latitude at the equator was 56,768 toises, La Condamine's 56,763 and Juan and Ulloa's 56, 768. Their measurements confirmed what had become accepted: the Earth was indeed flattened at the poles and that the Newtonian prediction was correct.

BY THAT TIME the Cassinis had already conceded and agreed that the Earth was a flattened sphere, the reason being another calculation from the measurement of the Paris Meridian. Cassini de Thury (Cassini III, Fig. 26) went back to basics, deciding to remeasure Picard's baseline (on which all the surveys depended) and to reduce

Fig. 26 César-François Cassini de Thury (Cassini III). © Observatoire de Paris

again all the measurements. Maupertuis had already done this, adding some subtleties in the calculation of star positions unknown to Picard. In two different attempts, he determined that the length of a degree between Paris and Amiens in the north was 56,926 toises and 57,183 toises: Picard had gotten 57,060. Cassini did something right at the basics of the issue though, remeasuring the baseline on the plain between Juvisy and Villejuif (in fact he measured a baseline that was not quite coincident but nearby, and corrected his measurement to find what Picard should have gotten).

Cassini found that Picard's baseline had been overestimated by 6 toises (12 meters too long over the 11 km). Cassini measured the baseline again, and again, and again with consistent results. This was a bombshell; Picard's baseline had been used as the fundamental length that entered into all the surveys of France. If the baseline had been overestimated then the size of the degree, the size of France, the size of the Earth- all of these were overestimated by the same factor and such a result would be very important, rippling through the whole field of geodesy. Its effect would enter into the discussions about the shape of the Earth and then into discussions about the law of gravity. And finally, it would affect his family's reputation. This was not a measurement to be regarded lightly and if he was sure that the error had been made, he would have to back it up when it came time to present this result to the Academy, his fellow scientists and the world. He called in three distinguished scientists from the Academy, Clairaut, Camus and Le Monnier, to witness the fifth measurement; it confirmed the previous four.

Cassini used the latest analysis by Maupertuis, corrected by this new calibration of Picard's baseline, to recalculate the scale of the degree getting 57,074 toises. The upshot of the reanalysis and the remeasurement was that Picard nearly had it correct, with two compensating errors, and in 1744, coupling the new measurement of the baseline with his own resurvey of the meridian, Cassini published La Méridienne Verifiée (Cassini 1744) on the lengths of the degree along the Paris Meridian. Some people have claimed that the work was ghost-written by his colleague Lacaille, and, if so, we can understand why he would not have wanted to go through the ordeal of writing what was to be a retraction of a deeply held view with which his family had become identified, but why he would want to appear to have fessed up to the reality. In this book he conceded– the length of a degree to the north of France was 57,081.5 toises and the length of a degree to the south of France was 57,048 toises. "Thus, according to these observations," wrote Cassini (or Lacaille), "degrees get smaller as one approached the Equator, which tends to prove that the Earth is flattened at the poles."

Length of one degree on the meridian

Place	Newton's calculation	Measurement	
Equator	56,637	56,768	Bouguer
		56,763	La Condamine
		56,768	Juan and Ulloa
France	57,048	57,060	Picard
		57,097	Cassini
Arctic Circle	57,332	57,438	Maupertuis
		57,196	Svanberg

Chapter 4
The Meridian and the Sun

At the time the astronomers of the Lapland expedition, including Maupertuis and Le Monnier, returned to Paris, builders were in the final stages to complete the Church of St. Sulpice. The Church has gained notoriety from its role in Dan Brown's bestseller, *The Da Vinci Code* (Chapter 9) and modern pilgrims can be seen, the novel under their arm, wandering around the nave. The Church authorities have even complained about the increase in visitors, who talk amongst themselves as they inspect this astronomical treasure. The noise disturbs the confessions in the south transept and the contemplation of those who wait their turn on the chairs nearby.

Places Church of St. Sulpice

The church stands near the center of Paris, on the left bank of the River Seine, north-west of the Palais de Luxembourg and about 200 meters west of the Paris Meridian. Its location is close to the meridian by chance, since the church was founded at the beginning of the 12th century well before the meridian was mapped out. It was a small church, several times enlarged between the 14th and 17th centuries, but in the first part of the 17th century it became too small to satisfy the pretensions of the inhabitants of the wealthy area that had grown around it (Terrien 2000). Starting in 1642 the original church was progressively razed to ground level and completely replaced by the present enormous church, the second largest (after Notre Dame) in Paris. The building of the church was accompanied by financial crisis so the building was completed in two stages: The first stage, which completed the eastern half of the church, lasted from 1646–1678 and then stalled (held up by financial problems). The second stage was completed between 1719 and 1732, with the façade completed in 1776. The proportions of the St. Sulpice Church and its solid classical columns are thought by some to be rather graceless but importantly it contains two unequalled treasures of art and science, namely a set of three murals by the French romantic painter Eugène Delacroix (1798–1863), who lived nearby, and the *meridiana* by Pierre Le Monnier.

The church is the home to a *meridiana*, built by Pierre Le Monnier (Gotteland and Camus 1997, Heilbron 2001, Kiner 2005, Rougé 2006). *Meridiana* is Italian for sundial, used in the English language for a scientific instrument that observes the passage of the Sun across the meridian. The *meridiana* works on principles drawn from nature; there are a million natural examples in copses, woods and forests. Every sunny day the Sun produces dappling of the ground beneath the trees under the canopy of leaves. The mass of leaves contains numerous small gaps, each of which acts as a pinhole camera and each hole can image the Sun on the ground, the ground acting as the image plane in the camera. The image plane is angled off

P. Murdin, *Full Meridian of Glory*,
DOI: 10.1007/978-0-387-75534-2_4, © Springer Science+Business Media, LLC 2009

the line of sight to the Sun because the ground is usually horizontal and the Sun is not usually directly overhead. The dappling under trees is thus a collection of elliptical images of the Sun.

As the Sun moves in the sky, rising towards its highest point at midday and then setting, the marks move on the ground below the trees. (Everybody knows that even if they are in the shade of a tree when they fall asleep after a picnic, they are in danger of having their siesta terminated by the light of the Sun as it moves the shadow of the tree off the sleeper). In the natural case it is hard to follow one particular image from a given pinhole because other branches intervene and block the image of the Sun that it produces. A *meridiana,* though, has one hole making one image, and a lens may be placed over the hole to give the image of the Sun extra clarity. At midday local time the image of the Sun lies on the meridian of the location in question. At midday on midsummer's day, when the Sun reaches its greatest elevation, its image is cast on the meridian at a point most nearly directly below the hole. By contrast on midwinter's day when shadows are longest the image is cast furthest from the point immediately below the hole.

Every day there are two measurements that can be made in a *meridiana* – the time at which the Sun crosses the meridian line and the position along the meridian line at which it does so. This constitutes a measurement of the position of the Sun at a particular moment. The *meridiana* can thus be used to check a theory of motion of the Sun, which attempts to say where the Sun is at any given time; this was the purpose of the St. Sulpice *meridiana*.

THE RE-BUILDING of St. Sulpice, including the *meridiana,* was completed by the Curé of St. Sulpice, Jean-Baptiste-Joseph Languet. For nearly half a century the building program had stalled and its re-activation was a tribute to Languet's fund-raising skills; it is said that he never returned from dinner with his wealthy parishioners without a silver service, and his collection was melted down to make a statue of the Virgin.

It was Languet who was responsible for the installation of the St. Sulpice *meridiana* as part of the new church. The accuracy of a *meridiana* depends upon the accuracy of the construction of the instrument, with the meridian line accurately north-south on a horizontal floor. Languet arranged for the floor of the transept, along which the meridian line is laid out, to be supported by its own pillars, independent of the rest, so that as the massive walls of the church settled the floor remained horizontal.

Languet's motive for making the *meridiana* was to determine better the dates for church celebrations, particularly Easter. The Council of Nicaea in 325 AD had the ambition to unite the Christian community across the world in its celebration of the events surrounding the crucifixion and therefore Easter was defined as the Sunday after the first Full Moon after the vernal equinox. To plan the celebrations, though, it is necessary to forecast this day, but this requires some astronomical knowledge about the motion of the Sun. If this knowledge is lacking, the predictions will be incorrect. The *meridiana* provided a means to check the predictions and to identify corrections to the basic formulae, which had been first developed in antiquity but as the centuries passed proved to be deficient.

In 1727, Languet commissioned an English clockmaker residing in Paris, Henry Sully (1680-1728), to make the original *meridiana*. Sully had come to France to establish a watch making industry at Versailles and "Frenchified" his name to Henri de Sully. Sully's motive for accepting the commission from Languet was different from Languet's motive for issuing it. The clocks of the churches of Paris were not synchronized and it was not unusual then to hear the noon bell ringing from different churches for half an hour or more. As a clockmaker who had been interested in the problem of determining longitude through the construction of an accurate chronometer, Sully was affronted by this lack of precision. He had in mind to use the *meridiana* to control the bell of St. Sulpice, to which the other bells would be synchronized. He was a member of a benevolent organization founded by the Comte de Clermont, the Société des Arts, to develop scientific and technical applications and saw civil advantages in correcting the irregular cacophony of hourly bells in the French capital. How could the citizens of Paris carry out their business efficiently, for example gather for a meeting, if they did not know the right time to act? The Society of Arts supported Sully in his campaign to provide a uniform time for the whole city. Sully completed plans for the construction of the *meridiana* and started to build it in 1727 but sadly died the year afterwards, before he could complete it.

The completion of Sully's project fell to Le Monnier. Starting in 1742, Le Monnier took up the challenge to complete the *meridiana* as a means to determine subtleties in the mathematics. He directed the instrument maker Claude Langlois in the engineering work (the same man that made quadrants and sectors for the expedition to Lapland that Le Monnier had used). Following Sully's footsteps, Le Monnier installed the entrance aperture for the *meridiana* in the southern window of the transept, and in fact there seems to have been more than one aperture since at certain times of the year the Sun cannot be seen from parts of the meridian line. For a few days around midsummer a ledge blocks the almost vertical beam of sunlight coming through the principal aperture from reaching the ground immediately below. Therefore there was a subsidiary aperture higher up above the ledge producing an image of the Sun on those days. The principal entrance aperture was a three-inch lens mounted on the right (west) edge of the plain window high (25.98 meters) above the south entrance, where there is now an empty hole in a mounting plate.

LE MONNIER'S MERIDIAN is marked on the floor of the church as a thin brass strip about 40 meters long and 7 mm wide with gradations marked alongside at intervals (Fig. 27). The strip is inlaid in white marble stone slabs about 20 cm wide and lies some 45 cm to the west of Sully's meridian. Sully's meridian was inscribed directly onto the stone floor but has mostly now has all been rubbed away. Faint lines associated with it, and the numerals "200" and "10" can be discerned beside the more identifiable meridian of Le Monnier in the northern transept and near the porch of the southern transept.

Le Monnier's meridian runs across the transept at an angle of 11° to the geometric axis of the church. In principle, European churches are built on a common plan and are shaped like a cross. The head of the cross is also the head of the church, where the altar and other holiest places are. The long axis of a church lies east-west, perpendicular to

Fig. 27 Le Monnier's meridian is marked as a brass strip (immediate foreground) which runs across the southern arm of the transept of the Curch of St. Sulpice and passes inside the altar railings (curving from the left foreground of this figure across the middle distance). It runs across an elliptical brass plate (foreground) that marks the image of the Sun at the spring equinox. It continues across the floor of the northern arm of the transept and runs up the northern wall. It is marked on a so-called gnomon, seen here in the distance. Photo by the author

the meridian on which the church stands. Thus, in principle the transept, or cross-piece of the cross, runs north-south along the meridian. The original 12th century church of St. Sulpice (whose foundations are in the crypt) was indeed accurately oriented to the cardinal points. The enlarged 17th century church was rotated slightly when it was rebuilt, in order to fit within the square and buildings surrounding St. Sulpice.

The meridian extends from the south, below the transept window and under the entrance aperture. The confessional is there as well, and people sit on chairs nearby, waiting their turn, brooding over their sins (or, perhaps, inventing imaginary ones that they can talk about). Occasionally the priest comes forth from the confessional in a flurry of priestly robes to rebuke noisy tourists disturbing the holiness of the moment. In front of the chairs where the penitents wait there is a plaque 90 cm square on the floor at the end of the brass meridian strip. It is engraved:

SOLSTITIUM ÆSTIUM
ANNI MDCC XLV
[P]RO NUTIATIONE AXIOS TERREN
OBLIQUITATE ECLIPTICÆ.

(The summer solstice in the year 1745. To find nutation, the terrestrial axis and the obliquity of the ecliptic).

There is a brass protective plate to protect this flagstone, but it has been under repair for the last 40 years; as a result the writing on the flagstone is becoming worn.

Where the line passes across the center of the church, just inside the altar gates at the crossing of the nave and the transept, a brass ellipse marks the equinoctial position of the Sun's image. There used to be a brass ellipse to mark the summer solstice but it has since been lost.

Though the Church of St. Sulpice is large, the transept is not wide enough to accommodate the whole of the meridian line on the floor where the image of the midday Sun moves towards winter solstice. The north end of the meridian continues up the wall in a vertical brass inlaid line on the bisecting axis of a 10 meter high stone pillar, or *gnomon*, made of red, gray and white marble.

Places The Gnomon of St. Sulpice

The instrument of St. Suplice is known in the Church as a "gnomon" or an instrument of knowledge. We would presently reserve the word for the part of a sundial that casts a shadow on to a scale to tell time. The *meridiana* of St. Sulpice has an erect pillar, like a gnomon, but it does not cast a shadow, it receives one. The gnomon stands against the north wall of the transept in a gloomy corner (Fig. 27). There is a horizontal arrow and two wavy lines part way (approx. 6 meters) up the pillar, marking the position of the Sun as it passes into the zodiacal signs of Aquarius and Sagittarius. Right at the top is the barely visible sign for Capricorn which marks the position of the solar image at midwinter solstice. The gnomon tapers upwards above the base pedestal from two brass crocodiles and floral florishes and is surmounted by a brass orb and cross. It was realized by Jean Nicholas Servandoni (1695-1766), French decorator, artist and scene-painter. He had designed the building of the façade of the church, though work on it was suspended after objections to its design, and it was completed by the architect Jean-François Chalgrin (1739–1811), also known for the Opéra.

The base (Fig. 28) has a long inscription. On the left (west) half of the base it states:

GNOMON ASTRONOMICUS
Ad Certam Paschalis
Æquinoctii Explorationem

(An astronomical gnomon for the determination of the Easter Equinox).

There follows an engraving of scientific instruments set on leafy branches: a book, globe, dividers, a square, armillary sphere, telescope, protractor, etc. These instruments are those that would have been used by Le Monnier to lay out the *meridiana*. There is also a snake curling through the leaves and around the measuring scale, presumably a warning by Servandoni against presuming too much with the knowledge being gained by the *meridiana*.

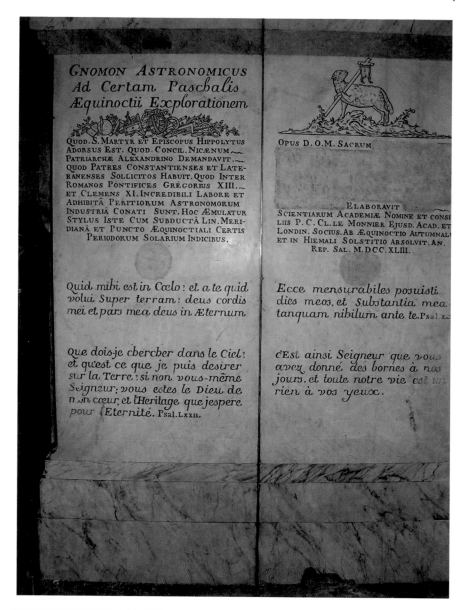

GNOMON ASTRONOMICUS
Ad Certam Paschalis
Æquinoctii Explorationem

QUOD. S. MARTYR ET EPISCOPUS HIPPOLYTUS
ADORSUS EST. QUOD. CONCIL. NICÆNUM
PATRIARCHÆ ALEXANDRINO DEMANDAVIT.
QUOD PATRES CONSTANTIENSES ET LATE-
RANENSES SOLLICITOS HABUIT. QUOD INTER
ROMANOS PONTIFICES GRÉGORIUS XIII.
ET CLEMENS XI. INCREDIBILI LABORE ET
ADHIBITÂ PERITIORUM ASTRONOMORUM
INDUSTRIÂ CONATI SUNT. HOC ÆMULATUR
STYLUS ISTE CUM SUBDUCTÂ LIN. MERI-
DIANÂ ET PUNCTO ÆQUINOCTIALI CERTIS
PERIODORUM SOLARIUM INDICIBUS.

OPUS D.O.M. SACRUM

ELABORAVIT
SCIENTIARUM ACADEMIÆ NOMINE ET CONSI
LIIS P.C. CL. LE MONNIER EJUSD. ACAD. ET
LONDIN. SOCIUS. AB ÆQUINOCTIO AUTUMNALI
ET IN HIEMALI SOLSTITIO ABSOLVIT. AN.
REP. SAL. M. DCC. XLIII.

Quid mihi est in Cœlo: et a te quid
volui Super terram: deus cordis
mei et pars mea, deus in Æternum.

Que dois-je chercher dans le Ciel:
et qu'est ce que je puis desirer
sur la Terre: si non vous-même
Seigneur; vous estes le Dieu de
mon cœur, et l'Heritage que jespere
pour l'Eternité. Psal. LXXII.

Ecce mensurabiles posuisti
dies meos, et Substantia mea
tanquam nibilum ante te. Psal. x...

c'Est ainsi Seigneur que vous
avez donné des bornes à nos
jours, et toute notre vie est...
rien à vos yeux.

Fig. 28 The base of the gnomon has a long inscription describing its purpose and history, as well as verses from the Psalms. Part of the inscription has been neatly excised by French revolutionaries who removed references to royalty and aristocrats. They respected the scientific function of the instrument, and left untouched the picture of the Lamb of God (right) and the decoration (left) of astronomical and mathematical instruments, intertwined with foliage and a snake

QUOD. S. MARTYR ET EPISCOPUS HIPPOLYTUS
ADORSUS EST. QUOD. CONCIL. NICÆNUM~
PATRIARCHÆ ALEXANDRINO DEMANAVIT.~
QUOD PATRES CONSTANTIENSES ET LATE-
RANENSES SOLLICITOS HABUIT. QUOD INTER
ROMANOS PONTIFICES GRÉGORIUS XII.~
ET CLEMENS XI. INCREDIBILI LABORE ET
ADHABITÃ PERITIORUM ASTRONOMORUM
INDUSTRIÃ CONATI SUNT. HOC ÆMULATUR
STYLUS ISTE CUM SUBDUCTÃ LIN. MERI-
DIANÃ ET PUNCTO ÆQUINOCTIALI CERTS
PERIDIORUM SOLARIUM INDICIBUS.

(*Because the holy martyr and bishop Hippolytus undertook the task, because the Council of Nicea entrusted it to the Patriarch of Alexandria, because the Constantine and Laterine fathers took care of it, because between the Roman pontiffs Gregory XII and Clement XI they attempted it with incredible work by skilled astronomers. This is emulated by this stylus with the meridian line drawn from certain evidence of the incidence of the Sun at the equinoctial point*).

[There follows a decoration that has been erased – some symbol of royalty, perhaps, or the zodiacal symbol for Pisces, according to Rougé (2006).]

Quid mihi est in Coelo? et a te quid
volui Super terram? deus cordis
mei et pars mea deus in Æternum

Que dois-je chercher dans le Ciel?
et qu'est ce que je puis desirer
sur la Terre? si non vous-même
Seigneur; vous estes le Dieu de
mon coeur, et l'Heritage que jespere
pour l'Eternité. Psal. LXXII.

(*Whom have I in heaven but thee? and there is none upon Earth that I desire before thee. My flesh and my heart faileth: but God is the strength of my heart, and my portion forever*).

The obelisk shows Latin and French versions of the same verse from Psalm 72 of the Roman Catholic bible, that is Psalm 73 in the Anglican version, verses 25–6, as translated here in the Authorised Version of the Bible, translated and published under King James in 1611.

On the east (left) half of the base:

OPUS D.O.M. SACRUM

[Several lines have here been deliberately but neatly obliterated. According to a drawing by the astronomer La Caille the wording was *REGIIS AUSPICII LUDOVICI XV, IN HANC BASILICAM MUNIFICI, FAVORE PRAESIDOIQUE D.D. FRED, PHILIPPEAUX, COMITIS DE MAUREPOIS, REGNI ADMINISTRI, EJUSDEM TEMPLI AEDITUI PRINCIPIS, NECNON D PHILIP; ORRI, REGNI ADMINISTRI, REGIORUM AERARII AEDIFICORUM PRAEFECTI PRIMARII.]ELABORAVIT* [the word *REGIAE* has been obliterated] *SCIENTIARUM ACADEMIÆ NOMINE ET CONSI-LIIS P.C.CL. LE MONNIER EJUSD. ACAD. ET LONDIN. SOCIUS. AB ÆQUINOCTIO AUTUMNALI, ET IN HIMALI SOLSTITIO ABSOLVIT. AN. REP. SAL. MDCCXLIII.*

(To God best and greatest, [King Louis XV, benefactor of this church, and under the administration of Jean-Frédéric Phélypeaux, the Comte de Maurepas, Minister of State, and Philibert Orry, Minister of State, Director General of the Kings Buildings] made this sacred work in the name of the [Royal] Academy of Sciences, and with the advice of Pierre-Charles Le Monnier of the same Academy and also a fellow of the [Royal] Society of London. He determined it from the autumn equinox and winter solstice in the year 1743.)

[Another decoration has been erased, the zodiacal sign of Scorpio according to Rougé (2006).]

Ecce mensurabiles posuisti
dies meos et Substantia mea
tanquam nibilum ante te. Psal. XXXVIII

C'Est ainsi Seigneur que vous
avez donné des bornes à nos
jours et toute notre vie est un
rien à vos yeux.

[Again Latin and French versions of Psalm 39, verse 5.] *(Behold, thou hast made my days as an handbreadth; and mine age is as nothing before thee.)*

The pillar is sometimes obscured by a jumble of piled up chairs, but if you can slip through them you can read the long inscription on the gnomon's base. Revolutionaries have neatly excised from it the names of the King, his aristocratic ministers and his royal symbols. The removal of the names of the royalists occurred during the period from 1793, when an official decree declared the suppression of Christianity, on the basis that the Catholic Church was an institution of privilege and its priests were associated too closely with the aristocrats. In place of Christianity, the revolutionaries instituted the religion of Deism, focused on a *Supreme Being*, who was thought to protect France. The church was taken over as a Temple of Reason and became a house of worship for the Supreme Being. A plaque above the main entrance of the church installed at that time declares that "The French people recognize the Supreme Being and the immortality of the soul." Later still, in 1801, it became a Temple of Victory; the statues and paintings were removed, the tombs desecrated, and the main altar destroyed. However, two pharmacists of the parish convinced the revolutionaries that the *meridiana* was a scientific instrument and should be preserved. Thus the gnomon was conserved (with the careful alterations to the inscription) and the raised altar area surrounded by its balustrade was left untouched, ultimately thanks to the portion of the meridian line that lay within.

Le Monnier observed the solstices at the Church between 1743 and 1791, an astonishing half century of consistent scientific observations, and plotted the position of the Sun on the meridian line throughout those nearly fifty years to better than a millimeter. Le Monnier later worked at an observatory in Rue St-Honoré run by the Capuchin order of monks and equipped with a wall mounted telescopes to observe the positions of stars. He was able to confirm by his observations of the Sun and the stars the obliquity of the Earth's axis, or its tilt. He found that the angle was decreasing at the rate of about 45 arc seconds per century (the modern value is 46.85 seconds per century). This decrease at the present time is part of a longer term periodic wobbling of the Earth's spinning axis.

Chapter 5
The Revolution and the Meter

Cassini IV became Director of the Paris Observatory in 1784 (Howard-Duff 1984). Louis XVI was then the king of France, Marie-Antoinette of Austria his consort. Marie-Antoinette was a beautiful, but capricious, queen and greatly disliked by the French people. By contrast, Louis XVI was at first a popular, but weak, king ruling France from his palace at Versailles at a time of ineffectively resolved political struggle. The struggle began as an argument about how to incorporate the bourgeoisie of rich merchants, traders and peasants (the "Third Estate") into the national decision-making process, joining the clergy (the "First Estate") and the nobility (the "Second Estate"). These latter two groups, the *ancien régime* (or *the old order*), were unwilling to share or restrict their feudal privileges, including rights to vote on the amount and kind of taxes and who should pay them, privileges that reflected social class, rather than empowerment from the franchise. Louis was faced with an increasingly severe national financial crisis which had its origins in the Seven Years War and in his own indiscreet extravagance. Wanting to reform and extend the taxation system, he called together representatives of the three Estates into a consultation called the Estates General, a body in which the three Estates had equal voting rights therefore ensuring that the Third Estate could always be outvoted by the other two, protecting their position.

The political struggle came to a head in the summer of 1789 when the Third Estate walked out of the Estates General and, in what became known as the Tennis Court Oath (its customary meeting place had been locked), established itself as the National Assembly, the national decision-making body. It declared that sovereignty lay not in the king but in the people; class distinctions were to be suppressed and it would implement a system of voting by head (effectively, representation by those who were being taxed). The king attempted to annul the declaration and ordered the military to enforce his decision against the National Assembly but the army refused. France increasingly became the scene of popular unrest, and a political conflict between the government and the privileged classes developed into a revolutionary conflict between the privileged classes and the people, with the king caught in the middle as arbiter and becoming increasingly out of touch with both sides.

The king ordered foreign regiments to the alert in Paris. This proved to be a mistake, and hearing of the potential interference by foreign armies in their home affairs, the Parisian mob marched on 14 July, 1789 to the Bastille, a fortress used

P. Murdin, *Full Meridian of Glory*,
DOI: 10.1007/978-0-387-75534-2_5, © Springer Science+Business Media, LLC 2009

to house political prisoners and a hated symbol of the *ancien régime*. The mob stormed the Bastille and captured it, as well as the arms that it contained. The Revolution had begun and the king was taken from the palace at Versailles to the palace in central Paris called the Tuileries, where he was kept under guard. Bastille Day (July 14th) continues to be celebrated by the French as the birthday of French national liberties.

Two days after the fall of the Bastille, the Observatory was caught up in the events. Suspicious of anything that they did not understand, the revolutionary mob invaded the Observatory to search it. Cassini was obliged to show the hundreds of invaders through the buildings. They did not find the gunpowder, weapons and food that they were looking for, and instead they stripped the roof of lead to make musket shot. This was the start of increasing Revolutionary involvement in the affairs of the Observatory.

The position of the scientists (or *savants*, to use the then current French word) in the Revolution was ambiguous.

On the one hand, the rise of science during the seventeenth and eighteenth centuries brought new thinking and criticism of the backwardness of monarchy and the Church. Diderot and D'Alembert began their project to create an encyclopaedia of human knowledge (*L'Encyclopédie*) in 1728, and had completed 30 volumes by the 1770's. The book was censored by the State in order to protect the status quo but contained enough criticism to set the scene for the French Revolution. Thus, many savants championed the liberal cause during the Revolution and even held office and others helped to further the aims of the Revolution in its agenda for reform based on rational and scientific thinking.

On the other hand, savants were often of wealthy, even aristocratic, ancestry and were subject to the same suspicions that were attached to others in a similar class. Some of them had, or were accused of having, counter-revolutionary views, resulting in purging from their institutions, imprisonment and even execution by guillotine.

Ideas The *Encyclopédie, ou dictionnaire raisonné des sciences, des arts et des métiers* ("Encyclopedia, or Reasoned Dictionary of the Sciences, Arts, and Crafts") (1751–1832)

The *Encyclopédie* originated as a translation of Ephraim Chambers's *Cyclopaedia*, first published in London in 1728, and the first encyclopaedia in the modern sense of a comprehensive, logically arranged collection of cross-referenced and indexed factual articles. Disputes over the rights and the position of one of the *Encyclopédie*'s first editors led to the *Encyclopédie* becoming a completely new project. Eventually it comprised 28 volumes and 71,818 articles. The editor in chief was Denis Diderot (1713–1784) a writer and philosopher who wrote the articles on economics, philosophy, politics, and religion, among others. He was assisted by Jean Le Rond d'Alembert (1717–1783) a mathematician, physicist and philosopher, who wrote the mathematical and scientific articles. Among its 96 authors were Baron d'Holbach (chemistry, mineralogy, politics, religion), Jean-Jacques Rousseau (music, political theory) and Voltaire (history, literature, philosophy). The *Encyclopédie* was intended to destroy superstition and elevate the role of rational knowledge. It was a complete summary of the thought in what has become known as the Age of the Enlightenment and controversial because of its religious tolerance, uplifting Protestant thought and challenging Catholic dogma. It played an important role as the intellectual underpinning of the French Revolution.

On the basis of its *Declaration of the Rights of Man* (26 August, 1789) the Assembly founded the new régime and its constitution of equality before the law and universal suffrage (the right to vote given to men, at least, without attention paid to social class). Now that France was free, the Revolutionary Government set out to make equality a reality by wholesale reform of French institutions, regulating brotherhood and the relationships between French people (*liberté, egalité, fraternité*). Local dialects were banned in favor of French, a principle which remains enshrined in the modern French constitution through the statement "the language of the Republic is French." Old regions were amalgamated and carved up into a system of Departments (*départments*), roughly equivalent to counties in the UK or the USA. (The Departments still survive as local administrative units; they have been numbered alphabetically since 1860. The numbers act as their abbreviations as the post- or zip-code in every French address and as the place of registration of every French car as proclaimed until recently on its licence plates).

As an essential part of its program of administrative reforms, the Revolutionary Government set out to make trade fair and uniform throughout the whole country by constructing a unified system of money and measures. How could trade be fair if the farmers who sold wheat to bakers in bushels bought bread back to eat in bushels of a different (smaller) size? In 1790, under Charles Tallyrand (1758–1838) and as part of the reforms, the National Assembly charged the Academy to design a new system of units to define weights, measures and other units. The new measurements were intended as a means for spreading enlightenment and fraternity among all the French people. The Academy decided that the system should consist of measuring units based on invariable quantities in nature. In the words of Condorcet while introducing the meter to the Legislative Assembly, they would be universal measures for all times and for all men.

People The Marquis of Condorcet (1743–1794)

Marie-Jean-Antoine-Nicolas de Caritat (the Marquis de Condorcet) was a mathematician who worked on the integral calculus and probability. Through his work on election statistics and voting methods, he was the founder of the modern science of Sociology. He became Secretary of the Academy of Sciences in 1774 and Secretary of the Legislative Assembly in 1791. He was a moderate who was arrested under orders by Robespierre in 1794, dying in prison in mysterious circumstances a few days later, most likely by suicide.

The intention of using natural units like the meter was to replace the arbitrary definitions. Some of the commonly-used units of length, for example, were defined in terms of something that was not fixed or permanent, or was mostly unavailable for comparison. Lengths based on the human body are obviously not well enough defined to avoid argument – a merchant selling cloth could hire people with small feet to measure length, so as to be able to claim that the cloth was as large a number of feet long as he could get away with. If the units of length were defined in terms of one particular man's thumb or feet, like the king's, he was unlikely to be available to settle a dispute, so it did not really help when the inch or the foot (originally the width of any thumb or the length of any foot) were replaced by the width of the king's thumb or the king's foot. The French *pouce* (thumb) and *pied-de-roi* (foot-of-the-king) were roughly equivalent to the British inch and foot but they were not standardized. They varied throughout France, with the Parisian *pied-de-roi* 11 percent longer than the one in Strasbourg and 10 percent shorter than the one in Bordeaux. This was not the end of the proliferation of units, which were as abundant as in Britain, sometimes even more so. For example, cloth was sold by the *aune* (ell) and land by the *perche* (rod).

There were two additional principles of the metric system. One was that all units other than the base units should be derived from these base units. The *pouce* and the *pied-du-roi* had been defined quite independently one from the other and it was only later that the foot and the inch were related by the convention that 12 inches make a foot. The second principle of the metric system was that multiples and sub-multiples of the units should be decimal. For a time there was a debate as to whether the sub-multiples should be twelfths. Peasants and trades people could halve or quarter lengths of cloth or an amount of produce by folding or dividing it. Twelfths made this easy but tenths were much more difficult; one can see this even today in a French market where eggs are sold in boxes of ten (made as two rows of five). If you want fewer eggs you can have four or six but not five because this means splitting the box lengthwise or stepwise and the seller will shrug, refusing to try.

The principles laid out by the Academy still underpin the modern metric system (now called the Système International d'Unités, abbreviated as the SI system). France created world-wide interest with this development and it resulted in 15 countries subscribing to the Convention on the Meter in 1875. The USA and Britain maintain their own everyday units of measurement, such as feet and inches, but define them in metric terms, and the SI system is used for all scientific and most but not all engineering applications. Where there is no universal common standard there is the risk of confusion, and in 1999 this risk became evident with the loss of the *Mars Climate Orbiter* spacecraft (Fig. 29). NASA lost the $125 million space-craft because a Lockheed Martin engineering team used English units of measure-ment to make it while the NASA used the SI system for spacecraft operation. The relationship between the two sets of documentation describing the manufacture and the operation of the spacecraft went awry. After a 286-day journey from Earth to Mars, the spacecraft fired its engine to push itself into orbit around Mars. The engine fired but the confusion in units meant that the wrong thrust was applied. Ultimately the spacecraft came much closer to Mars than planned and, after entering the dense parts of the atmosphere, bounced off into space and was lost.

A hundred years before the metric system was discussed by the eighteenth century scientists of the Academy, there had been debate as to the basis of the standard of length. In 1660, the Royal Society in London suggested that the standard should be a "new yard" or the length of a pendulum that swung uniformly back and forth each in one second (period 2 seconds). This would have led to a standard of about 39.2 inches, a few millimeters shorter than the meter used now. The proposal was supported by Isaac Newton. In 1675 the Italian polymath, Tito Livio Burattini (1617–1681), promoted this suggestion. He referred to this unit of length as a *mettro cattolico*, the first suggestion of the name *meter* (meaning *measure*) with the adjective contrasting the proposed stand-ard for use in Catholic countries against English units like the original yard.

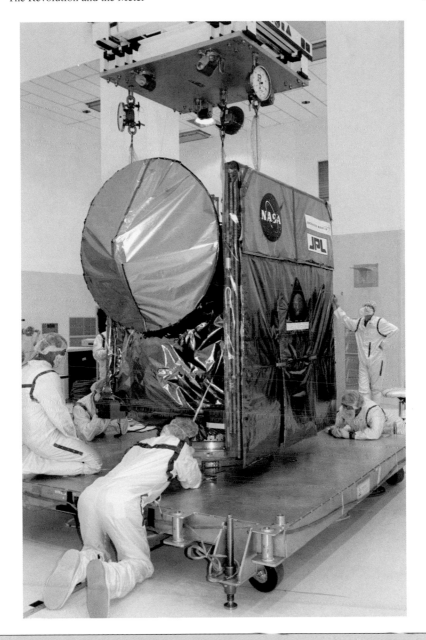

Fig. 29 The ill-fated Mars Climate Orbiter spacecraft is adjusted by technical staff in the clean room where it was assembled before its launch. It never reached Mars because of a muddle between the engineering units with which it was manufactured and the scientific, metric units with which it was operated. Photo copyright 1998, Lockheed Martin

When the Academy discussed the standard of length that was to be used by France, the length of a seconds pendulum was at first the main contender. The Academicians understood better than most the variation in gravity that there was between equator and pole because of the departure of the Earth from an ideal spherical shape and its rotation. They stipulated that the pendulum would have to be located at a latitude halfway between the equator and the pole and Cassini IV supported this proposal.

There was however an alternative school of thought dating back a hundred years; the French astronomer Gabriel Mouton (1618–1694) had the idea in 1670 of using part of the circumference of the Earth as the standard. He proposed the use of the length on the Earth's circumference of a minute of latitude as a standard, which he called a *mille*.[15]

The nautical mile was originally defined in a similar way as the length of 1 arc minute of latitude although it is now defined as exactly 1852 meters (the speed of 1 knot is 1 nautical mile per hour). Mouton proposed that the *mille* should be divided decimally in tenths, hundredths, etc. which he called the *centuria, decuria, virga, virgula, decima, centesima,* and the *millesima.* This decimal scheme, which greatly facilitated calculations, had been proposed in 1585 by the Flemish scientist Simon Stevin (1548–1621). Eventually it was decided to use Tito Burattini's name for a meter, but based on Gabriel Mouton's idea of using the Earth as a basis of its length, and to divide it decimally as suggested by Simon Stevin.

The National Assembly decided in 1791 that the measurements of the Paris Meridian from the equator to the pole should serve as the basis for the meter. Short of making the measurement by going to the equator and the North Pole, the meter was defined as the ten millionth part of 90° of latitude as pro-rated from a measurement of the extent of latitude along the Paris Meridian from Dunkerque to Barcelona. This meridian was centered on 45 degrees of latitude, neither at the equatorial bulge nor at the polar flattening, and it was already well reconnoitered. It was also international (running across the Franco-Spanish border) and the French Government invited international participation. This initiative to internationalize the metric system backfired. No country was happy to cooperate with the Revolutionary Government, seeing the project as one of French nationalism initiated by a government that had seized power by revolution and had executed thousands of the French relatives of people who held political power in other European countries. Spain participated, but only half-heatedly. Spain passively allowed the meridian to be surveyed in its own territory by the French scientists.

THE ACADEMY DECIDED, however, that the earlier surveys were not accurate enough to define the meter (Alder 2002). A new measuring device had become available in France making it possible to measure the circumference of the Earth to unprecedented accuracy and define the meter as a modern standard. This was the repeating circle first invented by the German mathematician and astronomer Johann

---

[15] *Mille* is *mile*, but the same word in French is *thousand*. The length of one mile originated as 1000 paces marched by the Roman legions.

Tobias Mayer (1723–1762) who was known for his calculations of the orbit of the Moon, making it possible to determine longitude at sea to one degree (about 100 km accuracy). The instrument was brought to a sophisticated and practical state by Jean-Charles Borda, whose version is known as the Borda Repeating Circle (Fig. 30).

Like a quadrant, the repeating circle measured the angle between two points, *A* and *B* (two stars, say, or the spires of two churches). The instrument consisted of two sighting telescopes independently mounted on an axis fixed to a stand, with the angle between them read on a circular scale. Unlike a quadrant which used one telescope and a rotating mirror system to sight on the two points and measuring the

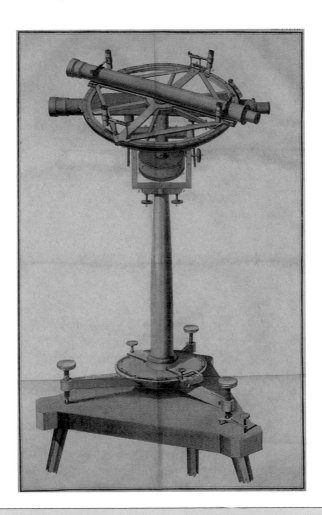

Fig. 30 The Borda Repeating Circle consisted of two telescopes mounted on a circular scale. The pedestal stood on a base that could be leveled accurately. The circular scale could be tilted so that the telescopes could view two target points (stars or geographical features) and the scale could be read with microscopes to show the angle between them

angle between them over 45°[16], the repeating circle had two independent telescopes and a scale that was a complete circle. It required two operators and was too big for one person to hold; it was more accurate than a conventional quadrant but less convenient to use. Mayer's idea was tried out in Britain by the Royal Navy, but because it was less practical its use was not adopted – the Navy's push to find a more compact instrument (easier to hold and operate on a ship in uncertain weather) caused the development of the sextant (a 60° angle on the scale). On land and in a scientific or a mapping survey the repeating circle was better.

People Jean-Charles Borda (1733–99)

Borda declined opportunities to become a Jesuit and started a career as a mathematician in the army where he studied ballistics and geometry. On the basis of his work he was made a member of the Academy in 1756. He saw action in the army in the Seven Years War and moved on to the navy, sailing extensively in the Atlantic Ocean between 1765 and 1782. At first Borda was on missions of exploration but later as a naval commander helped orchestrate the French victory over the British navy in support of the American War of Independence. In 1782 he was captured by the British and returned to France under parole, retiring to his estate during the early years of the French Revolution but returned to Paris to chair the Commission on Weights and Measures. Here, Borda proposed the decimalization of angles (a right angle of 100 degrees, each divided into 100 minutes, and each minute into 100 seconds) and calculated trigonometric tables based on decimal degree inputs.

If the repeating circle was used for astronomy, the scale was adjusted to be parallel to the line between the stars; if the instrument was used for surveying the scale was horizontal. A surveyor aligned one telescope on the first object, A, and clamped it on to the axis. Another surveyor aligned the other telescope on B, and clamped it accordingly. The scale read the angle, say θ, between the objects, but this was not the end of the measurement. The second surveyor rotated the two telescopes, both locked in place, passing his telescope to the first surveyor so that the second telescope sighted A. He unclamped the second telescope and passed it back to the second surveyor to sight B. The angle between the two telescopes was now 2θ and could therefore be more accurately measured. The doubling could be repeated perhaps ten or twenty times, progressively reducing the sighting error by repetition and by measuring the angles progressively against different sectors of the circle, reducing any systematic error in the engraving of the scale. The new instrument could read angles perhaps five or ten times more accurately than previously possible.

It is not clear whether the Academy selected Borda's repeating circle to measure the Paris Meridian because of its accuracy or whether Borda pressed the Academy to adopt it for the prestigious project in order to reflect well on his invention. Either way, the Academy commissioned four instruments to Borda's design (one survives and is housed in the museum of the Marseille Observatory) commissioned from the instrument maker Étienne Lenoir.

---

[16] The maximum angle that a quadrant could measure between two stars was double this because of the mirror reflection, hence the name "quadrant", a quarter of a circle being 90°.

People Étienne Lenoir (1744–1825)

Lenoir was an instrument maker, the son of a "poor but honest" stone mason, born in a village in the Loire valley and first trained as a locksmith. In 1772 he was employed as an instrument maker in Paris and studied mathematics by attending one of the free courses available to craftsmen. He set up his own business, supplying specialized astronomical instruments of high quality to leading scientists as well as making mathematical instruments for a larger market.

Lenoir played an important part in making and perfecting Borda's instrument (Turner 1989). He had just moved to Paris and was less than forty years old, working within (and triumphing) over a craft system in Paris that was restrictive and class ridden but progressively being broken down by reformers (Daumas 1972). In seventeenth-century France, craftsmen and workshops had been strictly regulated by guilds which a craftsman could enter under a very restrictive set of rules. In some cases the rules had nothing to do with talent at the job. Someone could become a guild member by being the son of a guild member, through election by the existing members, or by "privilege" (having residence in certain places, typically church property). In Lenoir's case, however, he did serve an apprenticeship and make a "masterpiece" as a demonstration of skill. After paying admittance fees and annual dues the craftsman could set up as a master craftsman with his own workshop but what he could make and the materials and tools that he could use were regulated by the guilds.

There was a dispute in the mid seventeenth-century over whether the making of mathematical instruments of copper was the prerogative of the cutlers (who made not only cutlery but also surgical instruments and cases for astrological instruments) or of the founders, who had the right to cast copper. The dispute was decided in favor of the founders and as a result mathematical instrument makers were in the same guild as people who cast guns, printer's type, and bells, and therefore regulated accordingly. The instrument makers were allowed to use a variety of tools but could only affix optical telescopes to the instruments, not make them, their lenses or mirrors. Guilds had the right to search workshops and to confiscate or destroy tools, patterns or products that were outside the regulations.

By the mid eighteenth-century the guild system was being liberalized but too slowly for the astronomers of the Academy. A patent system was set up by which an inventor could obtain a licence to manufacture his invention regardless of the guilds' restrictions and government commissions were supposed to be in the same class. A special class of instrument maker called *Ingénieur de Roi* (Engineer to the King) was created to recognize the most skilled craftsmen or those who had special commissions and exempt them from restrictions. Nevertheless in 1785 instruments commissioned by Cassini IV for the Paris Observatory were seized from Lenoir's workshop. Lenoir was not a master-craftsman but had a certificate of immunity from the police saying that he was the only instrument maker in Paris capable of the work. Unfortunately his work was confiscated by jurymen from the master Founders' Guild. Only by the intervention of Cassini was Lenoir able regain possession.

Cassini became increasingly frustrated by these battles and his difficulty in obtaining instruments of the right standard from French instrument makes. He set up a body of certified astronomical instrument makers, with Lenoir among the first seven appointed to the grade in 1788. He also established a new workshop at the Observatory; the existing commercially-run workshops were too unskilled and refused to make the unique astronomical instruments required by Cassini because there was little chance of a repeat order that would help defray the considerable capital sums necessary to set up the patterns and tools.

None of this progress could overcome the considerable class barrier between astronomer and instrument maker in France. English instrument makers like James Short and George Graham were scientists themselves, admitted to the Royal Society and treated as equals by their compatriot scientists. The optician John Dolland was even awarded the prestigious Copley Medal of the Royal Society. All of these English instrument makers wrote scientific papers that were published in the journals of the Royal Society. By contrast, Cassini treated Lenoir as the semi-literate worker that he was allowed to be, however experienced, ingenious and skilled he was, even though Cassini also wrote with great respect to Jesse Ramsden, an English instrument maker who made similar astronomical instruments to Lenoir.

The French instrument makers were on par with their English counterparts in their professional skill but nowhere near their equal in social standing. It took the events of 1792 to elevate their status when the Revolutionaries intervened in the elitist attitudes of the Academy, ordered everyone to *tutoyer*[17] their colleagues and to place the instrument makers on the decision-making committees. This change in status is illustrated by the part that Lenoir played, equally with Borda, in developing the repeating circle. Without his skill and contribution it would not have reached a practical implementation. Borda sought in vain for a workman capable of making the instrument with the required accuracy, especially the scale, until he discovered Lenoir. Even then he initially planned to have Lenoir construct the instrument but not the scales, intending to send them to England to be graduated. "In 1783 Borda asked M. Lenoir to construct these circles for him with the reservation that he would have them divided as he thought fit. The latter, stung by noble emulation, replied with laudable pride that no one other than himself would graduate the instruments that he had executed," wrote Jumelin in a report to the government in 1792.

It was Lenoir who thought of using two telescopes, one for balance on either side of the graduated circle and improved the mounting and thus the instrument's alignment and stability as well as the fineness of the adjustments that could be made to align the telescopes on their target. He made many examples of various sizes for the use of surveyors, hydrographers, navigators, and military engineers beyond the few made for the meridian survey and astronomy – one belonging to the Museum of the Conservatoire National des Arts et Métiers (National Academy of Arts and Sciences) in Paris bears the serial number 350, so there must have been a lot of them made.

THE ACADEMY had concluded that it was worth measuring the Paris Meridian again. It has to be said that there was a measure of self interest in the Academy's

---

[17] The French language has two forms of the word "you". The word *vous* is used both as a plural and as a respectful form of address; the word *tu* is used to a friend and equal (or inferior). To use the familiar form is called *tutoyer*.

decision. The budget allocated to define the metric system was three times the annual budget of the Academy, "a little *gateau* [cake]" according to Jean-Paul Marat, that the academicians could share. Marat sneered at the academicians, coining the scornful word *scientifiques* to replace the word in common use at the time, *savants* (wise people, intellectuals). This was to belittle what they did, in the same way that, in modern times, religious fundamentalists use *scientism* instead of *science*. But, just as the modern cosmologist Fred Hoyle was unhappy that his coinage of the phrase *Big Bang* became the accepted term for the present-day theory of the origin of the Universe (which Hoyle thought a completely erroneous concept), Marat would have been unhappy to see his coinage become the accepted word for *science* and *scientists*.

People Jean-Paul Marat (1743–1793)

Originally a doctor, Marat was the greatest revolutionary journalist of his day, twice an exiled, vitriolic editor of *l'Ami du Peuple* (*The People's Friend*) which was the preferred newspaper of the revolutionaries. He was made a Deputy in the National Convention in 1792 and helped stir up the class hatred that led (after his death) to the period known as The Terror in which 17,000 people were executed. He was himself assassinated, stabbed in his bath by a royalist sympathizer named Charlotte Corday, in an episode painted by Marat's colleague in the Convention Jacques-Louis David (1748–1825) in a picture called *The Death of Marat*. Today, this painting is displayed in the Musées Royeaux in Brussels – there are several copies, notably in the museums of Dijon, Reims, and Versailles.

However, the project to remeasure the Earth along the Paris Meridian was not simply a gravy train for the geodesists. Perhaps there was an element of this in the project; it could also have been a way that the members of the Academy could protect themselves from the charges of elitism being flung from the Revolutionaries in their direction by carrying out a project of practical use[18]. Additionally, there were inconsistencies between the measurements of the figure of the Earth in Italy and in France, through Rome and through Paris respectively. The discrepancies had to be identified and a new survey produced.

The Academy nominated Pierre-François-André Méchain (1744–1804) and Cassini IV to remeasure the Paris Meridian. Méchain would measure the southern stretch of the meridian into Spain. Cassini IV was to be, by inheritance, the leader of the project and should measure the northern meridian himself. Cassini dithered because he was unable to come to terms with the new political realties of revolutionary France and unwilling to leave Paris. He wished to direct an assistant to carry out the fieldwork but the Academy refused; he must measure for himself, the better to understand the data. Cassini would not do it. The Academy chose to replace him and appointed Jean-Baptiste-Joseph Delambre (1749–1822) in his place (Fig. 31).

According to historian I. Bernard Cohen (1970), "Delambre's early life resembles those novels of the nineteenth century in which industry overcomes hardship and is rewarded with social distinction and financial gain." He won a scholarship from his local school in Amiens to go to Paris and while there became skilled in languages.

---

[18] On a vist to the Beijing Observatory, I was told that during the Cultural Revolution in China in the 1960's the astronomers researched on sunspots and solar physics for the same reason.

Fig. 31 Jean-Baptiste-Joseph Delambre painted by Coroenne in 1879, after a bust held at the Observatoire de Paris. He leans on the volumes that describe the Basis of the Metric System (the first volume, propped up by a standard 1 kilogram weight, is open to display the title page) and stands in front of a terrestrial globe on which we seen Europe and Africa at the longitude of the Paris meridian. He picks off the length of a degree with a pair of dividers against a scale rule. © Observatoire de Paris

He was so poor that he lived for a year on only bread and water but became a tutor and to make himself more in demand taught himself mathematics and astronomy. In Paris he attended lectures by Joseph-Jérôme Lalande (1732–1807), a friend of Voltaire and self-titled the "most famous astronomer in the Universe." Delambre attracted his attention by reciting from memory a relevant passage on the constellations from the Greek poet Aratus and became Lalande's assistant and collaborator. Lalande was a militant atheist, well connected to powerful people in the revolutionary government and an excellent patron to have at this time. Delambre wrote an account of his experiences in carrying out his work on the Paris Meridian during what turned out to be the turmoil of the French Revolution, making light of the dangers with a dry and understated style (Méchain and Delambre 1806–10). Their adventures are the subject of a historical novel *The Measure of the World* by Denis Guedj (2001).

Like Delambre, Pierre Méchain too had humble origins (Fig. 32). He was the son of a ceiling plasterer but his mathematical ability attracted the attention of local patrons. His education in Paris was interrupted by poverty and he therefore became a tutor to two young noblemen. A story said to originate from Méchain himself says that while he was working as a tutor calamity struck. His father lost a crippling

Fig. 32 Pierre-François-André Méchain painted by Hurle in 1882. Dressed as befits his high status in the French scientific world, Méchain is interrupted at his work, but with his forefinger marks his place in a volume of Base of the Metric System. He does not meet our gaze. © Observatoire de Paris

lawsuit, and the son loyally agreed to sell his instruments to pay off the family debt. This setback proved his first stroke of fortune. His instruments were purchased by Jérôme Lalande. In this way he came into contact with Lalande who sent him the proofs of his book on astronomy. In 1772 Lalande procured for Méchain a job as hydrographer at the naval map archives in Versailles. The archives were transferred to Paris and Méchain began to draw up military maps of Germany and northern Italy. He also became active as an astronomical observer and eventually discovered several comets as well as several nebulae now known best by their listing in Messier's Catalogue. In 1785 he became the editor of the *Connaissance de Temps*, the French nautical almanac.

In 1787 Méchain had been one part of an Anglo-French team measuring the difference of longitude between Greenwich and Paris. His work with traditional surveying instruments was duplicated by Cassini and Legendre using Borda's new repeating circle and proved the superiority of the new device (although it took two people to work it). It was also a matter of pride that the repeating circle, invented and constructed by their countrymen, proved to be as accurate as the theodolite used by the English team and constructed by the acknowledged English master instrument-builder, Jesse Ramsden.

AUTHORIZATION TO BEGIN the new survey of the Paris Meridian arrived in June 1792 and Delambre immediately began preparatory work. Within two months he had established bases in Paris and was ready for the first night measurements. He sent an assistant, Michel Lefrançais de Lalande (Lalande's nephew) to the high point in Montmartre with instructions to light a signal on 10 August, 1792 on which he could sight his measuring telescope. There was no signal to be seen. What had gone wrong? Once again the Revolution had come between the scientists and their work. What had happened was the following.

The Revolutionary government had been cohabiting uneasily with King Louis XVI. His position was, to say the least, difficult, but had he been a wiser man or more surefooted politically, France might have turned out to be a constitutional monarchy like Britain rather than the republic it became. Ultimately it was the King who provoked the crisis, turning progressively for support to foreign relatives including those of his wife Marie-Antoinette, known as *"l'Autrichienne"* (the Austrian woman), and particularly Leopold II of Austria, Marie-Antoinette's brother. Reluctant to take part in the ceremonies and proclamations which consolidated the new constitution, King Louis XVI and the royal family plotted escape. The King played it cool, participating in normal kingly business up to the last minute. On 19 June 1791, in the Tuileries Palace, he even received the Committee on Weights and Measures – Borda, Cassini, Coulomb, Lavoisier, Legendre and others. He exchanged pleasantries with Cassini, and then asked him: "M. Cassini, I have been told that you will measure the Paris Meridian again. Your father and grandfather already did this before you. Do you think you can do it better than they did?" Cassini replied, "Sire, I would not boast of it if I had not a great advantage over them. The instruments my father and grandfather used only gave the measure of an angle to within 15 arc seconds. The Chevalier de Borda has devised an instrument that will measure angles to one arc second. This will be to my credit." The very next day, as he must have known was the plan while engaging in this chit-chat, Louis XVI and a large entourage fled from Paris in a carriage specially made in secret for the purpose (Tackett 2003). At Varennes, a few miles from the Austrian border, the party was stopped to have their papers inspected. Something aroused the suspicion of the locals (one story is that the King's profile was recognized from a coin), and the party was asked to wait. A local judge also recognized him and indiscreetly bent his knee in acknowledgement. The party was arrested and was brought back to Paris under close guard.

The incident increased suspicion of the King and his motives in fleeing to a foreign country. He had left behind a declaration against any form of constitutional monarchy, which he had previously supported. The implication was that he was going into self-imposed exile, to rally France against the revolution and return triumphant when it was defeated. Indeed, by the summer of 1792 Austria and Prussian armies were heading into France and towards Paris in order to rescue and restore the monarchy. Convinced that the King was behind the invasion, ten thousand revolutionaries stormed the Tuileries, killed hundreds of his guards and set the royal palace on fire. The King fled for protection into the Assembly but was handed over and taken into custody. As a result, the King was deposed and the Republic of France was proclaimed on 25 August, 1792.

Delambre had arranged with Lefrançais that they should start the measurements between the Observatory and Montmartre on 10 August, 1792, the day that turned out to be that of the storming of the Tuileries. In making their plans, Delambre wrote (Méchain & Delambre 1806–10), "we knew nothing of the one thing and another that was happening at the Tuileries." In those circumstances to light signal flares from the most prominent hill in Paris would have certainly attracted attention and may well have invited death at the hands of the mob; even to make the journey was dangerous. Lefrançais gave up any idea of traveling out to Montmartre to light the signals that evening and indeed would not have been permitted to leave Paris to do so.

In subsequent days Delambre tried to survey alternative locations around Paris, but it had proved too difficult for him to carry out any kind of work in and around the city. Fearful of putting a foot wrong, officials refused to act without permits, and the authorities in the capital were fearful of issuing them. Delambre had to look for an alternative place to start the surveying measurements and settled on the Château de Belle-Assise, near Lagny to the north-east of Paris. In September 1792, Delambre traveled there in his carriage, one of two (the other for Méchain) specially designed and built by Borda as mobile laboratories. There was a folding table on which to lay out papers and maps, and in addition the seats were convertible to beds if (as was often the case) it was necessary to sleep at a remote station where there was no accommodation available. Also, there were storage compartments on the walls for scientific instruments and long compartments on the ceiling for rolls of maps and a large trunk on the back for the repeating circle.

With the help of three assistants Delambre set up his observing station at Belle-Assise. He had completed three days of observations and was gathering his equipment together to leave when he was interrupted by the local militia. Counter-revolutionaries were gathering for action, the four men were strangers and, looking out with telescopes from the towers of the château, they might be spies – so thought the locals. Indeed, Delambre was recognized by one of them as the person who had a few days earlier been seeking permission to light signal fires in the city. He was taken in a downpour across fields to Lagny where he produced his papers. They had, however, been signed by King Louis XVI who was no longer king. Far from establishing Delambre's legitimacy, the papers increased suspicion. Delambre explained that his task was to measure the world and that he was a member of the Academy. "There is no more 'cademy," said one of the militia, "We are all equal now." Delambre and his colleagues spent the night in jail until the militia received confirmation of the legitimacy of their mission.

Delambre finished his work at Belle-Assise and set off back to Paris. Barricades had been set up at the villages to stop aristocrats from fleeing the vengeance of the revolutionaries and the embrace of "Madame Guillotine." The men were stopped at each village, examined, and then allowed to continue until they reached Epinay-sur-Seine when they were again detained by the local militia. The telescopes attracted unwanted attention, and as usual it seemed that telescopes were seen as useful for spying and were not documented clearly on the passes and permits.

Delambre tried to explain his mission to measure the Earth but this story sounded incredible to the gathering crowd. He was taken into St-Denis to account for himself to the authorities. Meanwhile, a thousand people were assembled in the square, and Delambre was escorted through them to cries of "Long Live France! Down with the aristocrats!" Delambre tried to explain what he was doing (Méchain & Delambre 1806–10):

> The instruments were spread out on the square and I was obliged to recommence my lecture on geodesy, the first lessons of which I had given earlier that day in Epinay. I was not heard any more favorably this time. The day was coming to a close. It was increasingly difficult to see. My audience was quite large. The front rows heard without understanding. Others, further back, heard less and saw nothing. Impatient murmurs began to be heard. A few voices proposed one of those expeditious methods, so much in use those days, which cut through all difficulties and put an end to all doubts.

The Chief Administrator of the district, Denis-Nicolas Noël, saved Delambre from the guillotine, taking him inside the town hall under pretext that he would be questioned more closely. Noël kept Delambre overnight and when the mob dispersed Delambre was allowed to complete his journey back to the capital.

Over the winter of 1792/93, Delambre prepared for the expedition into the north. As the winter weather ameliorated to spring, the Prussian army was massing on the French border intending to invade and restore the monarchy; the plains of Flanders were a battlefield, not for the first or last time. Delambre traveled to Dunkerque to start his survey, intending to work south and as he did so the revolution intensified. Led by Robespierre and the so-called Committee of Public Safety and ruling on the basis of rumor and prejudice, Robespierre and crew carried out its wishes by imprisonment and execution of dissenters and doubters.

Delambre, together with Borda, Lavoisier, Laplace, Coulomb, and Brisson, were purged from their positions on the Commission for Weights and Measures. Their offence had been to support the chemist Lavoisier, who had fallen foul of the

Committee of Public Safety. He was one of a score of farmers-general who had been taken prisoner and put on trial (the farmers-general had collected taxes on behalf of the king). Borda organized the drafting of a letter calling attention to Lavoisier's essential skills used in support of the work of the Commission for Weights and Measures and stating that it was urgent for him to return to the important work which was interrupted by his arrest. The letter was not well received and immediately called forth a decree, signed by Robespierre himself, and four others, noting "the importance of delegating duties and assigning various missions only to men whose republican zeal and hatred of kings make them worthy of the public trust." The Committee of Public Safety declared that Delambre, Borda, Lavoisier, Laplace, and Coulomb should "cease as of this day [23 December, 1793] to be members of the Commission for Weights and Measures and shall deliver forthwith, with an inventory, any instruments, calculations, notes and papers in their possession that may pertain to the measurement project." Delambre received his copy of the decree and had to return to Paris in midwinter 1793–74. His house had been sealed and he was able to enter only accompanied by local commissaries who scrutinized every paper in the house including Delambre's membership certificate from the Royal Society of London, written in Latin and signed by King George – a suspicious, royalist paper indeed! Delambre once again faced interrogation and the possibility of the guillotine.

MÉCHAIN HAD NO EASIER time of it in his work measuring the southern part of the meridian and his share of the proposed survey was half the length of Delambre's – the distance from Rodez to Barcelona versus Delambre's section that ran from Rodez to Dunkerque. The reason was that Delambre was resurveying what had already been measured twice whereas Méchain was to survey the entirely new sector in Spain and would have to reconnoiter for appropriate triangulation stations as well as measure them. "Little did we know," wrote Delambre (Méchain and Delambre 1806–10), referring to the civil unrest of the Revolution, "that the greatest difficulties would be found at the gates of Paris." Méchain left Paris on 24 June, 1792 accompanied by his servant Citizen Tranchot. At Essonne he was immediately detained by a roadblock of anxious citizens, armed and nervous, looking for counter-revolutionaries, but the municipal officials had not lost the respect of the people and he was allowed to continue his journey. Méchain arrived a week later in Perpignan and by August, accompanied by a Spanish officer, Captain Gonzales, he was hauling his equipment into the Pyrenees and looking for a triangulation route through Catalonia towards Barcelona. Gonzales, an officer in the Spanish Navy, had been invited to participate in the observations, to facilitate the relationships between Méchain and the local Spanish authorities, and no doubt to keep an eye on the two foreigners!

The region's inhabitants were not French revolutionaries or aristocrats, but worse. They were Spanish Bourbons, suspicious of French intentions, and Catalonian rebels, suspicious of anything to do with either of the Spanish and French governments and doubly suspicious of something to do with both! Possibly they were also smugglers or bandits, answerable to no-one except themselves. In the inns, the beds that Méchain and his party slept on were boards on trestles and the food they ate was all but inedible. By contrast with what else they faced as they journeyed this was merely discomfort; as they traveled through the border country they risked their lives to marauding bands of renegades and patriots. Nevertheless Méchain worked through the arc from Perpignan towards Barcelona, and by the end of 1792 he was established in a fortress location on the hill of Montjuïc (in French Mont-Jouy), overlooking Barcelona, where he decided to stay to make measurements of the latitude his most southern station of the Meridian survey. By accident on 10 January, 1793 he discovered a new comet; he explained he wasn't looking for it lest anyone think that he had been diverted from the main purpose of the observations.

Comets are thought by the superstitious to be harbingers of wars, famine and the death of kings. On 21 January, 1793 Louis XVI was executed, war between France and Spain followed, and France was plunged deeper into economic chaos.

Places Montjuïc

The hill of Montjuïc (428 m high) was fortified in the eleventh century to protect the port of Barcelona. The present Castillo de Montjuïc is an eighteenth century castle in the traditional star shape of continental fortifications, designed in 1751 by the French military architect Vaubon. It is reached by cable car from the funfair and is now a museum of militaria, with displays of model soldiers, uniforms and armaments overlooked by a large statue of General Franco on a horse. He may be reflecting on the use of the castle as a prison during the Spanish Civil War where his opponents were taken to be imprisoned, tortured and in many cases killed. In this he was continuing the long history of the fortress as a place of repression (four-hundred suspected anarchists and left wingers were incarcerated there in a round up after the attack on the Corpus Christi procession in Barcelona in 1896). If you can put these grim thoughts behind you, you can take pleasure in looking out to the splendid vistas over the city and visiting the Miró Museum, the Palau Nacional (Museum of Catalan Art) and others nearby. Méchain's geodetic observations are memorialized by the Montjuïc Castle geodesic marker.

Because Méchain was an alien, he was expelled from Montjuïc and restricted to the town of Barcelona. He had completed his observing work and perhaps at first enjoyed the respite, but his inactivity was further enforced by an accident in which he broke some ribs and his right arm. By the autumn of 1793, he was recovered enough and trusted enough to resume work on the Spanish side of the border but was still not permitted to return to France. He reduced his observations at Montjuïc and became troubled because the observations had an inconsistent result for the latitude of his mountain observing station (a discrepancy of 3 arc seconds, corresponding to about 100 meters). Over the winter of 1793–74, Méchain took up observations again from his hotel in Barcelona as a check, linking the two observing stations through a grudgingly issued one-day-pass to Montjuïc. To his horror the result clearly showed inconsistency but the data did not resolve what it was.

Méchain began to spiral into what we would now classify as depression. His scientific rigor had forced him to recognize that there was an error in his observations but he began to doubt that whether he was able to resolve it. In addition, his situation gave him cause for concern; as an alien in Spain his position was restricted and every day as the French armies penetrated deeper into Catalonia the people became less friendly. On the other hand, they would not let him leave to return to France given what he had seen of the defences of Barcelona, and if he stayed in Spain, would the authorities in France conclude that he had defected with who-knows-what consequences for his family back in Paris? Lacking the knowledge that everyone in Paris had come to the obvious conclusion that he was imprisoned in Spain, Méchain made arrangements to depart to neutral Italy, and left for Genoa in June.

BACK IN PARIS, Delambre remained under suspicion. He had returned to his house and was permitted to enter only when accompanied by local commissaries who scrutinized every paper in the house. The officials paid great attention to a certificate of his membership to the Royal Society of London and to papers covered with what might be calculations or enciphered secrets. At the Observatory, Cassini IV bitterly opposed the involvement of the new revolutionary government in observatory affairs and was affronted by the execution of King Louis XIV in January 1793. Cassini was betrayed by three assistant astronomers drunk on the revolution with the supposed equality of their knowledge of science to his. They demanded co-authorship of papers which they had all worked on under Cassini's direction and not mere acknowledgement.

The Observatory was reorganized on egalitarian lines. Cassini was demoted and his salary halved, while the three assistants were promoted to be at his level and rate of pay. One of them was made the first director under a new rotating arrangement for the position. In protest Cassini resigned, was ultimately evicted and his scientific work seized. When he protested again, he was imprisoned and faced with the suggestion by one of the three assistants (Alexandre Ruelle) that he be sent to trial by the Revolutionary Tribunal. Cassini escaped trial and what would have been the near certainty of the guillotine but was imprisoned for 8 months. He retired from scientific work to his château and the comparative safety of local politics. He spent his old age writing polemics justifying his own position and defending the scientific reputation of the family.

The three assistant astronomers, now Professors, remained working at the Observatory. Ruelle was discovered to have lied about some solar data, fraudulently concocted from theory and not from observations. Ruelle fell from favor and was imprisoned. The two remaining former assistants invited Delambre to join them as a Professor at the Observatory, but he wisely continued to stay out of the public eye remaining to work at his country manor.

In June 1795 the Observatory was put under the control of a newly created state body, the Bureau of Longitudes, and there was yet again another reorganization. With his impeccable republican credentials, Joseph-Jérôme Lalande was appointed Director. The meridian project was restarted in a calmer political atmosphere, and Delambre was asked to take it up again serving under the title of *Astronomer of the War Department*. When Delambre left the city in June 1795 and headed south to the most southern extent of his northern half of the meridian in Bourges, he must have felt glad to have kept his head and to have left behind the terror of the past two years in Paris.

Delambre still had practical difficulties to overcome, such as convincing innkeepers to accept the letters of credit that the Government had given him with which to pay for supplies, but he was able to contemplate completing his scientific work. He linked Bourges, Orléans and the Sologne region and headed back north to Dunkerque to measure its latitude and compare it with Méchain's measurement of the latitude of Montjuïc.

Places Bourges

The graceful gothic cathedral of Saint-Etienne at Bourges is distinguished by its 13th century stained glass windows, which rise to the roof among the five tiers of arches along the flanks of the cathedral. In 1795 Delambre used the pelican-shaped weather vane on the cathedral tower as a central triangulation site in his survey of the region. The cathedral has a *meridiana*, which was installed in 1757, three years before his death, by a canon of Notre-Dame des Salles. A thin copper strip on the pavement between the third and fourth side-arches marks the north-south line. The cathedral has no transept and the *meridiana* cuts across the nave at an angle of about 80° to the axis of the nave between the third and fourth arches. There are no markings along the strip except for a circle carved in the stone floor near the south end, presumably the position of the Sun at midsummer, so the *meridiana* was not very well developed as an astronomical instrument. Perhaps this is because the installation of the *meridiana* was so close to the death of its rather aged initiator. The aperture of the *meridiana* is a hole made in one of the high stained glass windows of the south side. The hole is in the elbow of the figure in green on the left of center, and it is rather surprising that such a hole could be made in one of these ancient windows. There is also a stone memorial in the floor beside the meridian strip at the south end commemorating its maker. It reads as follows:

DOM PETRUS ESTERLIN SUBDIAC
CAN. HUI ECCL. OBIIT DIE 2 ^
MARTII. 1760 ÆTAT 72

(Father Pierre Esterlin, sub-deacon and canon of this Church. Died 2 March, 1760 aged 72).

IN ITALY, MÉCHAIN learned of Delambre's fall from favor and concluded from the fact that he had not heard anything from France that he had been abandoned. In a way this was a relief because he would not have to explain the error of measurement of the latitude of Montjuïc, but the relief was short-lived when he learned that the project had been restarted and that he had been appointed to continue his part in its completion. The large increase in salary he was offered could not erase his despair at the thought of having to reveal his scientific incompetence to his colleagues. The disgrace seemed imminent when Méchain was called back to Paris for discussions about the project. No wonder he contemplated the words of refugees from France who advised about the possibly fatal uncertainties of a return to home; should he rather accept the less dangerous poverty of exile?

Places Lieusaint

The meridian passes near to the western boundary of the Seine et Marne department, where Lieusaint was the northern limit for the 1798 survey from this point to Melun by Jean-Baptiste Delambre. A stone pillar marks his triangulation point here.

Méchain oscillated between the call of duty and his family as well as the fear of conspiracy, enmity, exposure, and disgrace. Finally in April 1795 he returned on the mail boat to Marseille but even though he was back in France, he dithered, staying in Marseille for five months. Eventually he decided to go, not to Paris, but to Perpignan to take up the measurements again but he waited too long; the weather had turned and he could not make the measurements that he needed. Worse, Delambre, who had been working hard during the summer, was starting to take an interest in Méchain's measurements, and wanted to combine them into the final analysis. Delambre arranged with Méchain that they would continue the triangulation towards the city of Rodez, central in the unfinished sector of the meridian, and meet there. Méchain, therefore, would have to face up to his mistakes.

Delambre arrived in Rodez in August 1797, having finished his work, and was able to return to Paris but he had heard nothing from Méchain for most of the year. He contacted Méchain's wife at her accommodation at the Observatory where she had been living comfortably on Méchain's increased salary and was surprised to learn that her husband had not met Delambre. In the winter of 1797, Méchain confessed why- yes, he had some problems with the mountain weather and had not made as rapid progress as Delambre, but he had also been reducing his data and discovered further discrepancies, wanting to return to Barcelona for further observations. He wrote to Delambre in November: "In this situation I have chosen to remain in the horrific exile I have long bewailed, far from my other duties, far from all I hold dear, and far from my own best interests…Either I will soon recover the strength and energy I should never have lost, or I will soon cease to exist."

Delambre did not like the sound of this, taking it as an apparent reference to suicide. He would not be reassured by another letter in March 1798: "After all that has happened I can no longer show myself anywhere and my only wish is to be annihilated." Delambre persuaded Mme Méchain to visit her husband to reassure and persuade him to finish the work or to accept that Delambre would come to the south to help Méchain out.

IN THE SUMMER OF 1798, Méchain was surveying in southern France. Even several years after the most intense days of the Revolution, his work still aroused suspicion. A letter in the archives at Albi seeks information from local commissioners about suspicious devices erected by Méchain, one on a tower of a castle in Montredon and another in Montalet, both of them targets for his surveying instruments (the devices were suspiciously painted white, the royal color):

16 August, 1798

To the commissioners near the communes of Lacaune and Montalet

I am informed, citizen, that someone has erected a structure atop the highest point of the chateau of Montredon. In Lacaune, at the place known as Montalet, someone has erected a sort of machine, painted white, which looks a lot like a tent. I am told, moreover, that the plans for this machine were provided by a stranger claiming to come from Paris, and this same person directed the construction.

I urge you, citizen, to go immediately to the chateau of Montredon to ascertain whether this machine exists and to enquire of the owners of the chateau as to the purpose for which it was constructed and, further, to determine whether it may be of use to the enemies of the republic. You will kindly describe the structure to me and pass on any information that you may obtain, attached to this letter. It will then be easy for me to determine the truth.

In the present crisis no functionary serving as an agent of the government may take anything for granted. Even the simplest-seeming objects may conceal perverse intentions. You will therefore discharge this commission with all the zeal, scrupulousness and speed of which you are capable.

The target at Montalet was destroyed by hotheads several times and Méchain had to request that guards be posted. His request was passed down from commander to commander, but the importance of his request was recognized and fulfilled:

3 September, 1798

To the Commune of Lacaune

The adjutant general who commands the armed forces in the département was unable to provide the seven infantrymen we requested to protect the target in Montalet. We therefore did not feel obliged to replace them by a similar number of hussars, as the general suggested, in view of the likely difficulty of keeping the men supplied in these mountains. The importance of Citizen Méchain's operations should be kept in mind at all times, however. You will kindly requisition two or three men from the mobile column both to guard the target and to provide Citizen Méchain with whatever facilities and help he may require. You will billet the men as specified in our letter of the eleventh of this month because the aforementioned astronomer requires this target for observations he is to make in the canton. You will leave the guards at the target until we authorize you in writing to remove them.

If you deem it necessary, you may relieve these men with others. Do everything in your power to ensure that we are not again disobliged by the news that the operations of this commissioner of the government have been suspended owing to difficulties with the target in Montalet.

MME MÉCHAIN MET HER HUSBAND in Rodez in July 1798 for the first time in six years in order to persuade him to return to Paris and hand over his measurements. But even after talking with him for five weeks, in what must have been an emotional time, her mission was a failure: "I have begged him in vain to write to you," she wrote to Delambre in September 1798, "and to agree to measure the baseline of Perpignan with you. Not wishing to upset me he always answered vaguely. For the first time, my husband has dissimulated with me."

Delambre decided that he had to go to meet Méchain himself, regardless of the niceties of scientific collaboration as it was not polite to set up a joint project and then to muscle in on what was agreed to be the collaborator's involvement. But Delambre was running out of time because the National Convention had passed a law in April 1795 embedding the metric system in French life, setting up the prefixes (kilo, mega, etc.), the names of the units (meter, liter, etc.), and their definitions (in the case of the meter, a preliminary definition pending the completion of the geodesic survey).

As part of the process to internationalize the meter, the French Government had called an International Conference of scientists to agree on the metric system as the international scientific standard. This was the only way, reasoned the Academy, that it would be possible to gain acceptance for a meter based on French measurements along a part of France. The French government agreed, but with a different spin- the Conference would also promote the "glory" of the nations prepared to cooperate in the undertaking. This proviso would certainly exclude Britain since the country was France's main European rival with which, under Napoleon Bonaparte, France was actually at war, but also America which had been seeking rapprochement with its former colonial master. The Conference would be attended by France's political allies the Netherlands, Denmark, Switzerland, Spain, and Italy. British and American scientists were not invited. Nor were German scientists invited, in part because they would represent a country with fragmented political power that would be unable to agree on a trade standard and in part because the German scientists were firmly in favor of the pendulum as a standard of length.

The date of the International Conference had originally been set as September 1798, and although the foreign scientists forming the Commission had begun to assemble in

Paris in time, the Conference was postponed until November. Delambre had to have something to report, preferably side by side with Méchain. Delambre traveled to the cities of Narbonne and Carcassonne in the south of France, closing in on Méchain. Méchain persisted in refusing to meet him, making a succession of excuses but at last they met early November 1798. At first Méchain was adamant that he would not return to Paris and reveal the perceived inadequacy of his work. He needed more time to polish the work before exposing it to foreign scientists, stating, "I will not expose myself to this final humiliation." It was not as bad as Méchain thought, though, because the Bureau of Longitudes had agreed that Méchain would be the next Director of the Observatory. Whatever he thought of his own work, others respected his capability, and Méchain eventually agreed to return to Paris to attend the International Conference and to hand over the data. In mid-November 1798 Méchain and Delambre returned to the capital, to a reception given by government ministers and the entire Academy of Sciences for the visiting foreign scientists and the two French scientist-heroes who had taken the measure of France against the tribulations of the times.

The International Commission set up a competition between the two astronomers. They would each determine the latitude of Paris and use their measurements to calculate the meter separately in terms of the distance to Dunkerque and to Barcelona. The degree of agreement would demonstrate the accuracy that had been achieved. This was Méchain's nightmare come true; he was as unsure of his data as ever. While Delambre presented his data to the International Commission who went over it detail by detail and accepted the measurements, Méchain refused to hand anything over and the international scientists became restless. The Danish astronomer Thomas Brugge began to conclude that the whole thing was a sham and left for Copenhagen in January 1799. Méchain was ordered to hand over his data and agreed to provide an edited summary, which he did in March 1799. It must have been a relief to Méchain that the International Commission found everything in superb order. As later developments showed (below) he had successfully pulled the wool over their eyes.

The Commission's members individually set out to combine all the measurements and to calculate the final results which were stunningly unexpected. The Earth was certainly not an ideal sphere, that was already known from the earlier surveys by the Cassinis, but it was not a uniform ellipsoid either, not even any solid that could be created by revolving a curve about the Earth's axis. The flattening of the Earth was different, as measured in different segments of the Paris Meridian. The measurements revealed the irregularities of the Earth's surface and that not all meridians were the same.

The International Commission could not determine a consistent value for the overall flattening of the Earth so it decided to adopt the value from the longest survey of the Paris Meridian into Lapland, together with the measurements made in Peru, and to apply it to the new data to estimate the circumference of the Earth from equator to pole; from this followed the standard length of the meter.

FOR PRACTICAL PURPOSES, a platinum rod was produced that was supposed to merely represent the ideal, unchanging length of the meter because it was more accessible than the circumference of the Earth. It was presented to the French Assembly in June 1799 without a mention of the uncertainties in the natural standard from which it was derived, and copies were made and displayed publicly around Paris so that anyone could check a length of cloth or chain that they had bought.

Places Standard lengths

On the face of the Palais de Luxembourg opposite number 36 is mounted one of the 16 marble meters issued by the revolutionary government between February 1796 and December 1797 as a public reference standard, propagating the meter to practical use.

Following the French example, there is a similar device dating from 1876 mounted on the north wall of Trafalgar Square in London, labeled "Imperial Standards of Length, placed on this site by the Standards Department of the Board of Trade MDCCCLXXVI - Standards of Length at 62 degrees F." It marks out lengths of one foot, two feet, a yard, and an inch. Along the bottom of the wall is marked a standard chain (66 feet).

In practice the platinum rod itself was declared to be the standard meter, weakening the founding principle of the metric system that it should be based on natural reproducible measurements. To be sure, the speeches made to the Assembly reminded the legislators that the meter was based on the size of the Earth and made everyone a "co-owner of the world" – fine, democratic stuff. The very basis of the metric system, however, had turned into a political sham. Not until 1960 when the meter was redefined in terms of the wavelength of a particular spectral line of the element krypton did the metric system return to the ideal scientific principles with which it had started.

The 1960 definition was replaced in 1983 when the meter was defined as the distance that light will travel in a specified fraction of a second and where a second is defined in terms of a certain number of the oscillations of a caesium atom. Today, the meter has become a secondary standard and the standard unit is effectively the light-second.

The Paris Observatory maintains its connections with these definitions through the Bureau Internationale de l'Heure (International Office of Time), maintaining at the Observatory the world's standards of time and position by intercomparison of the world's atomic clocks and position-measuring telescopes even though time and longitude is conventionally reckoned as starting in Greenwich (see Section 13). To keep its own clocks stable and isothermal, they are housed in the basement of the building, its foundations extending as far below ground (27 m) as the building is high.

What happened to Delambre and Méchain, you may ask? Honors and accolades were showered upon both but Méchain was still severely depressed. He avoided visitors and buried himself in his scientific work as Director of the Paris Observatory; he had not changed and suffered from self-doubt and depression. In 1801 he called for more measurements that would extend the meridian past Barcelona into the Balearic Islands. The southern part of the measured arc would be distant from mountains and the irregularities in the Earth's surface would be revealed by the earlier survey, but the survey would give him the chance to repeat and supersede the worryingly discrepant measurements which he had made ten years before. Eventually the Bureau of Longitudes approved the plan but appointed another astronomer to lead it believing that Méchain was better employed as the director of the Observatory. Méchain insisted that he should do it himself.

IN 1803 MÉCHAIN again set off to work in the south of France and Spain. We can see something of his black depression at this time from the following letter to Delambre:

> Hell and all the plagues it spews upon the earth – storms, wars, pestilence and dark intrigues – have all been unleashed against me. What demon still awaits me? But vain exhortation will solve nothing, nor complete my task.

From Ibiza he found that he could not sight the mainland station at Montsia, as he had planned. Attempting to complete his task before he returned to Paris, Méchain was forced to change the pattern of his surveying and search for stations further south than he had intended. He did not complete the measurements nor return home. He was weakened by age, the work and his diet. He succumbed to a fever – malaria, which he caught during his expeditions to remote triangulation stations that he established in the marshy, mosquito ridden swamps of Albufera. He collapsed and in September 1804 died in Valencia. He had mercifully been released from his troubles. In his eulogy of Méchain, Delambre showed the deepest sympathy for his perfectionism:

> Never did he consider these observations, the most exact ever achieved in this domain and conducted with unsurpassed certainty and precision, and never did he consider them sufficiently perfected.

Delambre was appointed to write up the work on the Meridian, which he did in a work called *Bas du Système Métrique* (Foundation of the Metric System). This took him seven years and in compiling the definitive measurements he was scandalized and perplexed by the irregularities in Méchain's work. Delambre discovered how Méchain had altered observations to make them consistent, covering up discrepancies in the latitude of Montjuïc. Working on his own, head down in his office, refusing to discuss his work with anyone, providing only edited summaries to others, Méchain had fudged the data, a process that was facilitated by the way that he kept the original observations in pencil on scraps of data. Figures had been erased and altered; original papers had been "lost."

Delambre was heartened, though, by the discovery that Méchain had perpetuated his scientific fraud in an "honest" way. The averages had not been altered, but what we would now call the standard error, or spread of observations, had been reduced. This made the observations seem more consistent than they otherwise would have appeared. However, Delambre decided to suppress this discovery and explained why in notes deposited in the archives of the Paris Observatory, where they lay unread and under seal until they were found by historian Ken Alder two centuries later (Alder 2002).

> I have carefully silenced anything that might alter in the least the good reputation M. Méchain rightly enjoyed for the care he put into all his observations and calculations. If he dissimulated a few anomalous results which he feared would be blamed on his lack of care or skill, if he succumbed to the temptation to alter several series of observations…at least he did so in such a way that the altered data never entered into the calculation of the meridian.

Delambre completed the publication of the three-volume account of the work to re-measure the Paris Meridian and established the natural standard of the meter in 1810 twenty years after the project was started (Méchain and Delambre 1806–10). Delambre himself presented a copy to Napoleon, and in the margin of his own copy of the book Delambre noted the Emperor's response: "Conquests pass and such works remain." This wry remark both prophesied Napoleon's own eventual defeat and memorialized the enduring legacy of the science of the meridian project.

Chapter 6
The Paris Meridian in the Napoleonic Wars

Méchain's unfinished work on the survey of the Paris Meridian in the Balearic Islands was completed by Jean-Baptiste Biot and François Arago.

People Jean-Baptiste Biot (1774–1862)

> Biot was educated as a mathematician, with his father intending that he should be active in commerce, but after a spell in the army as a gunner during the revolutionary wars in 1792–94, he became a professor of mathematics. He was appointed to the Paris Observatory in 1806 and later became professor of astronomy at the Paris Faculty of Sciences of the University of France (established by Napoleon in 1808) where he spent the rest of his career. He was an ardent exponent of Newton's Theory of Gravity and his textbook was the one from which Sir George Airy, later the Astronomer Royal at Greenwich, learned astronomy. In 1803, Biot investigated the reports that stones had fallen from the sky at l'Aigle in the Orne. At the time such reports had been thought to be superstition and misguided but regardless Biot visited the area, spoke to witnesses and collected specimens. He showed that the composition of the stones bore no resemblance to the geology of the region, and his report was the beginning of the realization of the truth about meteorites- that they fall from space.

Dominique François Jean Arago (1786–1853) was born and educated in the south of France in Estagel, Roussillon, in the eastern Pyrenees and at that time a small village (Daumas 1943; see also Anon 1854 and Howard-Duff 1986). His father originated in Catalonia, Spain and was cashier at the regional mint. Arago was one of a family of eight children comprised of six sons and two daughters. As he grew to a tall and lean maturity (Fig. 33), he taught himself mathematics to a standard good enough to enter the examination in 1803 for places at the Polytechnic School in Perpignan and was attracted by the splendid uniform of an officer of engineers to a military career. The school had been founded only eight years earlier to provide opportunities for people to develop their skills for the service of the state.

But as Arago was about to set off, his father had a visitor, Pierre Méchain, who was carrying out his survey of the southernmost section of the Paris Meridian. Arago's father asked Méchain his advice regarding his son's career and the mathematical education that he was seeking. "With the frankness which is my characteristic," Méchain said, "I ought not to leave you unaware that it appears to me improbable that your son, left to himself, can have rendered himself completely master of the subjects of which the programme consists. If, however, he be admitted, let him

P. Murdin, *Full Meridian of Glory*,
DOI: 10.1007/978-0-387-75534-2_6, © Springer Science+Business Media, LLC 2009

Fig. 33 François Arago painted by Charles Steuben in 1832, aged 46. © Observatoire de Paris

be destined for the artillery or the engineers; the career of the sciences, of which you have talked to me, is really too difficult to go through..." It was with these discouraging words ringing in his ears that Arago took the entrance examination, unexpectedly coming in first place. He entered the Polytechnic School at the end of 1803, where, according to his immodestly drafted autobiography (*The History of My Youth*, Arago 1857) he became the head of his class.

Méchain died in Spain in September 1804 and his son, who was Secretary of the Paris Observatory, resigned. Jean-Baptiste Biot of the Observatory was an examiner of the Polytechnic School and had noticed Arago's work. So too had Denis Poisson, professor of the Polytechnic School, who recommended Arago to the Director of the Observatory to take Méchain's place. With his mathematical ability, Arago was invited to replace the younger Méchain as the Observatory Secretary. He did not want to give up a military career for one of science, especially after the discouraging words of Méchain senior, but nevertheless he took leave from the Polytechnic to take up the position on a trial basis. He worked with the telescopes in the Observatory as an observer, measuring star positions, and also began to work with Biot on the refraction of light through gases, which was a subject of practical importance in astronomy because of the way that star positions are distorted by the atmosphere. Part of the work included the accurate weighing of gases held in glass globes, and this important data was used by the chemist Joseph Louis Gay-Lussac (1778–1850) in the first steps to derive the atomic weights of carbon and hydrogen.

While Arago worked with Biot they discussed the value of completing the work in Spain of extending the Paris Meridian which had been interrupted by Méchain's death. Biot and Arago submitted a plan to complete the southern extension of the Paris Meridian to Pierre-Simon Laplace, who was then the Director of the Paris

Observatory. Laplace accepted the proposal by Biot and Arago and they set off to measure the southern sector of the Paris Meridian in the Mediterranean. Arago's part as a young man in making these measurements (at the start of which he celebrated his 20th birthday) seems like an adventure story; the work turned out to be more dangerous than it would have seemed, even by the standards of the age.

People Pierre-Simon Laplace (1749–1827)

A brilliant mathematician, Laplace had worked on differential equations and the stability of the solar system. He was the author of the *Nebula Hypothesis*, which envisioned that the solar system formed from condensations in a flat rotating nebula, a theory which foreshadowed the modern concept. In 1790 Laplace had been a member of the committee of the Academy of Sciences to standardize weights and measures, but along with Lavoisier, Borda, Coulomb, and Delambre, he had been thrown off because their Republican credentials were not good enough. Laplace lived 75 km from Paris and wisely stayed there during the Reign of Terror, thus escaping the guillotine, and in 1795 when the Academy was revived and established the Bureau de Longitudes, Laplace joined its staff and became director.

ARAGO AND BIOT set out for Spain from Paris early in 1806 and visited some of Méchain's stations on the way to re-measure some of his triangulations. In Spain they were joined by two Spanish Commissioners, Señores Chaix and Rodriguez, who brought with them the scientific instruments left by Méchain. Biot and Arago scaled one of the bleakest mountains near Valencia, the Desierto de las Palmas, where Méchain had created a triangulation base by levelling a small platform. They attempted to measure from the coast across the Mediterranean Sea to the Balearic Islands, where Chaix and Rodriguez climbed Mount Campvey. The light that had been installed during Méchain's expedition was difficult to locate and to see reliably because its optical system had been misaligned. The light on Mount Campvey was on the island of Ibiza, 150 km across the sea from Desierto de las Palmas on the coast of the Spanish peninsula, and Arago and Biot could not see it for months through the atmospheric absorption.

"It will be easily imagined what must be the ennui experienced by a young and active astronomer, confined to an elevated peak, having for his walk only a space of twenty square meters, and for diversion only the conversation of two Carthusians, whose monastery was situated at the foot of the mountain, and who came in secret, infringing the rule of their order," wrote Arago in his autobiography. The two monks came separately to talk with Arago and Biot, but, each one learning that the other had infringed the rule of silence, they both conspired not to report each other. The younger man was a reluctant monk, forced into the order by his family. In front of Arago and Biot he relayed anti-religious sentiment based on information he had gathered while taking confessions. Biot suspected him of being an agent-provocateur and sharply turned the conversation to a different topic. The next day, Arago had to persuade the monk to lay aside a gun with which he had murderous intentions towards Biot. The older monk was no more moral than the younger; Arago records his disgust and caution at being invited during mass conducted by this man to join him in drinking more wine than was part of the sacrament. The disgust rose from religious feelings of respect for the mass, the caution from a suspicion that the wine might be poisoned in revenge for Biot's rebukes.

After six months Arago and Biot located the triangulation point on the mountain of Campvey. They celebrated and only then did Biot show his companion a letter written by Méchain, found among his papers, in which he expressed his doubts that accurate triangulation to the Balearic Islands would be possible. "Even supposing that it is possible," he wrote, "the time it would take is so long that it could overwhelm me, kill me... This unhappy commission whose success is so prolonged, much more, so uncertain, will be more than likely my failure."

Arago's youthful exuberance and, one imagines, indiscretion, drew him into exhilarating adventures. During a tourist visit to Murviedro he met a young señorita who invited him to her grandmother's for refreshments. Upon leaving her house, Arago hired a mule-drawn carriage and the services of the mule-driver, Isidro, but unfortunately the carriage was attacked by her jealous fiancé and an accomplice, one of whom died when trampled by the mule and run over by the carriage wheels in the mêlée. The next day, wrote Arago in his own account, the Spanish Captain General told him that a man had been found crushed on the road to Murviedro: "I gave him an account of the prowess of Isidro's mule, and no more was said."

Arago conceived a plan to install a new station on a mountain near Cullera which was home to bandits; at this time, Spain was descending into anarchy and chaos through its insurrection against Napoleonic rulers. When Arago had taken up his post in Paris, Napoleon was at the height of his powers, crowned by Pope Pius VII as Emperor in the church of Notre Dame in Paris on 2 December, 1804. At barely thirty years old, he had become First Consul of France in November 1799 and de facto ruler of the First Republic. Napoleon set about suppressing the democratic institutions that had been created by the Revolution and developing the centralist principles which remain in some aspects of French government to this day. In 1804, he published the *Code Napoléon*, which swept away the remaining privileges of the *ancien régime* in the codification of French laws and increased the power of the French state. He created the gendarmerie as a para-military police force to suppress opposition and ruled some of the neighboring countries in the Empire directly from Paris (the Netherlands and the Low Countries, the Illyrian Province of the eastern Adriatic coast, western Italy and Catalonia in Spain). In others he installed family members as rulers – Westphalia in northern Germany, the Kingdoms of Italy and of Naples, and Spain.

Napoleon's power to the west of Europe was limited by Portugal and by the sea power of Britain, which won a historic victory over the French navy at Trafalgar (near Gibraltar) in 1805, but to the east his influence spread into modern Germany. In 1805 as Arago was planning his journey to Spain, Napoleon was expanding his Empire further east still with the defeat of the Russian and Austrian armies at Austerlitz. Prussia was defeated at Jena in 1806, and Russia was defeated again in February 1807 at Eylau, its army slaughtered. As a consequence Napoleon was able to create another puppet rule in Poland called the Grand Duchy of Warsaw.

In an attempt to bankrupt and starve his enemy, Napoleon denied the use of continental ports for trade with Britain. In retaliation the British Royal Navy blockaded French and other Napoleonic ports on the west coast of Europe and in the Mediterranean Sea, denying imports from friendly and neutral countries, which created as much hardship in France and Spain as did the denial of continental trade to Britain. Massive conscription of local people, burdensome taxation to support the war effort, brutal occupation by the French armies, as well as hunger and poverty brought unrest to Spain as the rule of law and order broke down (also called the "Spanish Ulcer").

Arago had to penetrate into the remoter countryside of Spain to seek out suitable mountaintops as triangulation points and such areas were controlled by brigands. His escort of national guards fought a battle to gain him the appropriate access to Cullera and as a result the brigands were dispersed into the countryside. Arago had to work overnight at the station and had a cabin built which he shared with his servant. He was sheltering there during a stormy, rainy night when there was a knock at the door; asked to identify himself the man claimed to be a customs officer. The servant opened the door and there entered "a magnificent man, armed to the teeth." Arago offered him hospitality and shelter from the weather. They all settled down to sleep overnight. On waking and chatting to the "customs officer" at breakfast time, Arago saw two more people approaching the cabin to pay a visit, as did the stranger, his eyes flashing as he recognized the mayor (*alcaid*) of Cullera and his principal constable (*alguazil*). This "customs officer" did not stay to greet them, but ran off, springing from rock to rock like a gazelle. He was the leader of the bandits, and according to Arago's account had chosen not to shoot at his two most cruel enemies out of gratitude for Arago's shelter and food.

Some days after this, Arago received another visit from the "customs officer" and allowed him shelter and sleep a second time. Arago's servant, an ex-soldier, was inclined to kill the visitor but Arago restrained him: "We are not discharging the duties of the police," explained Arago, "and if we are known to have harmed this man his followers would make our work impossible." In the morning, Arago told the bandit chief that he knew who he was and asked if they had anything to fear from the bandits. He candidly replied that they realized that Arago would not carry money to such a remote place and that it would bring too much trouble from the Spanish authorities if a French official under state protection was hurt. He added his gratitude for Arago's hospitality and again Arago was left unharmed and unrobbed. But on leaving the station, Arago encountered a second gang of bandits (not parties to any favorable understanding about his position) and was forced to take refuge, mules and all, in the kitchen of a farmhouse listening in the dark to the voices of the gang searching for him nearby.

Working their way still further south, Arago and Biot created a triangulation station on the mountain of Mongo near San Antonio, from where they were able to link to Formentera, one of the smaller Balearic Islands at nearly 100 km distance. Mongo dropped steeply on three sides to the sea and access there was difficult. They were able to gain the summit with the help of a party of sailors who built a shelter there, but winter was well advanced and the February cold at the summit was hard to bear.

Both men traveled to the islands in March 1807, but some of their measurements were frustrated by instrumental failures when the Borda repeating circle broke. They returned to the Spanish mainland and Biot then returned to Paris carrying the repeating circle for repairs. Arago used the remaining working equipment to repeat measurements along the coast, but his work was becoming increasingly difficult due to the progressive unrest. In Valencia he witnessed the outcome of a trial by the Inquisition accusing a woman as a witch. The sentence was that she should be stripped to the waist, covered with honey and feathers, and paraded through the city on the back of an ass, her face turned towards the tail. She was then struck several times on the back with a shovel. Arago was outraged by the superstitious spectacle taking place in what was one of the principal cities of Spain and the home of a long-established university.

In November 1807, Biot rejoined Arago at Valencia. They traveled back to Formentera in the Balearic Islands and spent the last weeks of the year making observations to determine the latitude of this, the most southerly point of the Paris Meridian to be surveyed. They also determined the length of a seconds pendulum (one that beats seconds) at this location. Biot left Arago in January 1808 and Arago embarked for Majorca and the mountain of Campvey, where he established a station on the Clop de Galazo. He succeeded in joining Majorca to Formentera, to Mongo and to Desierto de las Palmas in a large surveying grid, measuring the latitude at each place and thus determining the scale of a degree and a half of latitude in one survey.

BUT HIS STAY IN THE Balearic Islands was soon to come to an end. As Arago finished his scientific work on Majorca, the political and civil unrest in Spain was developing into a full-scale uprising, aided by Britain and Portugal, known as the Peninsular War (1808–1814) in which France lost 300,000 soldiers. Arago had to make his way home as tension in Napoleon's Empire was increasing, eventually leading to a terrible miscalculation when Napoleon invaded Russia in wintertime in 1812, losing some 400,000 soldiers in the invasion and the subsequent retreat from Moscow.

Attempting in 1807 to put down the Spanish insurrection, French armies had invaded Spain while Arago was working on the Clop de Galazo on Majorca. While Arago was making his observations early in 1808, Napoleon's generals conquered San Sebastian, Pamplona, Figueras, and Barcelona, as well as brutally suppressing an uprising in Madrid. King Charles IV of Spain abdicated in favor of his son Ferdinand, and Napoleon moved his brother Joseph from his position as King of Naples to become King of Spain. This attempt to crack down on the growing rebellion misfired and Spain burst into open revolution.

Arago's observing station on Majorca overlooked a strategic port and his activity making astronomical observations at night was taken by the local Spanish population to be signal making by a French spy to the French invading army. The local population began an uprising when M. Berthémie, a French Napoleonic officer, came to Majorca on a mission from Napoleon to conscript the Spanish islanders to go to Toulon to join with a French force. Berthémie was threatened with death at the hands of the mob, which he evaded by offering himself up for imprisonment by the Spanish authorities on the island. The mob then remembered Arago on his mountain and decided to settle with him instead. Arago gathered up the sheets of his observations and left his mountain base as the mob approached. As they crossed paths he shared friendly words with them in the local Majorcan dialect, Mallorquí, which is a dialect of Catalan, Arago's birth language, and they did not suspect that the man that they had encountered was the French official whom they were seeking.

Arago thus evaded capture and made his way to Palma de Majorca where, in June 1808, he too offered himself up for imprisonment by the Spanish authorities in the castle of the Captain General. There Arago had the unique experience of reading in a newspaper about his own execution, which was worrying, although he took some satisfaction in the account of the heroic way he was reported to have met his death as he was hung. It was clear, however, that his situation was serious, and he was concerned that the false account might presage the fact.

Arago and Berthémie made arrangements to leave Majorca and in July 1808 they chartered a boat and left the island. By August they had landed in Algiers and sought refuge in the house of the French consul, with whose aid they chartered another (Algerian) boat to sail to Marseille, disguised as merchants, one from Schwekat in Hungary and the other from Loeben in Austria. The boat was approaching Marseille in mid-August when it was captured by a Spanish corsair and taken to Rosas on the Spanish coast near the French border. Cross-examined on arrival, Arago was momentarily disconcerted that he had forgotten whether he was supposed to be the Hungarian or the Austrian but fortunately guessed correctly. At first he was imprisoned in a windmill and then in September and October he was transferred first to a fortress and then to the hulks at Palmos, where he was almost starved of food. On the intercession of the Dey of Algiers[19], Arago and Berthémie were eventually freed along with the Algerian crew and the boat and were allowed to continue their journey towards Marseille.

Within sight of Marseille in November 1808, the boat was caught up in a storm, the strong wind of the mistral propelling it on to the coast of Sardinia where it was severely damaged. It limped across the Mediterranean Sea south to Bejaia in Algeria for repairs where Arago and Berthémie joined a caravan to walk through hostile Arabs to Algiers, reaching the city on Christmas Day 1808 and the relative safety of the French consulate. Even in Algiers, outside the arena of the direct confrontation of the Big Powers, he was not safe. A new Dey of Algiers, replacing the beheaded Dey who had negotiated their freedom from Spanish corsairs, refused to allow Arago and Berthiers to leave. In February 1809, the Dey issued a demand for payment of back taxes from France of two or three hundred thousand French francs. When the French consul, on Napoleon's orders, refused to pay, all the French nationals in Algiers were imprisoned save for Arago who had arranged a bail bond through the Swedish consul. This Dey's reign was, however, limited and he was deposed and hung. The French consul was persuaded to pay the demand to the new, third Dey by the Jewish community resident in Algiers who wished to normalize trading relations between Algiers and France. Finally Arago was released to sail to France in June 1809.

---

[19]"Dey" was the title given to the Regent of Algiers (and the Regent of Tunis) during the Ottoman Empire from 1671 until the French conquered Algiers in 1830. Algiers was a city state, under the Ottoman Sultan and was the chief haven of the Barbary pirates. As a regent, the dey was an elected supreme ruler, chosen by local leaders to govern for life (unless deposed). The dey was advised by an inner cabinet of ministers called the "divan" – the word has passed into the English language for the seats on which they sat to deliberate.

As Arago sailed north and neared France on the 1 July, 1809 a British warship was blockading Marseille. Arago's ship was boarded, and his heart must have sunk at the thought of a further period of imprisonment. He wrote about this incident in his autobiography (Arago 1854, 1857):

> The vessel in which I was, although laden with bales of cotton, had some corsair papers of the regency [of Algiers], and was the reputed escort of three richly laden merchant vessels which were going to France,

> We were off Marseille on the 1st of July, when an English frigate came to stop our passage: "I will not take you", said the English captain, "but you will go towards the Hyère Islands, and Admiral Collingwood will decide your fate."

> "I have received," answered the Barbary captain, "an express commission to take these vessels to Marseille and I will execute it."

> "You individually can do what may seem to you to be best," answered the Englishman: "as to the merchant vessels under your escort, they will be, I repeat, taken to Admiral Collingwood." And he immediately gave orders to those vessels to sail to the East.

"Admiral Collingwood" was Admiral Cuthbert Collingwood, the Royal Navy's Commander-in-Chief of the Mediterranean in 1809 (Adams 2005), who commanded a squadron of 15 British warships that were blockading a French naval fleet of 21 ships (including six Russian) tied up in the harbor of Toulon under the command of Vice-Admiral Ganteaume. Collingwood's mission was primarily to stop this fleet from supporting Napoleon's armies, but also to prevent supplies from reaching France and to put pressure on the French government in its home. It was under these instructions that the British warship intercepted Arago's group of ships.

The ship that intercepted Arago has up to now not been identified. I believe that it was HMS *Minstrel* (Murdin 2005), a 423-ton sloop with 18 guns. At the time of the Napoleonic Wars, the word "sloop" covered a variety of warship types. They were captained by an officer with the rank of commander and had 10–18 guns, so *Minstrel* was at the upper end of this range; if the ship had been any larger it would have been a frigate. Its captain was Commander John Hollinworth, a man who entered the Royal Navy as a midshipman and was promoted first to Lieutenant in 1802 and then to commander in 1806, a year before he became the captain of the *Minstrel* (which he commanded from 1807–1809). In his final years he reached the rank of Rear Admiral, the lowest of the three ranks of admiral in the British Navy.

In the UK's National Archive I found Hollinworth's and the *Minstrel's* logs for 1809. The National Archive is located at the Public Record Office beside the River Thames in Kew, and this is where the government's records are kept, including the original logs of the ships of the Royal Navy. In the early nineteenth century, each Royal Navy ship kept several logs. The original ship's logs were hour-to-hour records made at the time of the incidents reported. Sometimes the handwriting (several different hands) is very difficult to read, and one can readily imagine the entries being written on the moving deck of a sailing ship. The captain's logs, however, are fair copies and filed with the Admiralty, the London headquarters of the Royal Navy, written at leisure in a neat hand using the ship's log as a reference.

Minstrel's log for 1809 (TNA reference ADM 53/820) is a large leather-bound ledger of plain pages covered with filthy dust. Its pages are frayed and water stained but speak eloquently of life in the Royal Navy. Like an adventurous seaman, the log has been active at sea in all weathers and like a seaman too old to fight, it ended its service days in the Admiralty, confined to an office ashore and spending its years shelved in retirement. Only occasionally, as in the case of my visit, does the log come alive again, recounting its stories.

At the time Arago arrived in Marseille, HMS *Minstrel* was blockading the approaches. Collingwood had orders to prevent supplies from reaching France in order to put pressure on the French government at its home. From time to time he dispatched individual ships to "cause annoyance to the enemy's trade." His British warships captured all the French-flagged merchant ships that they could and chased others until they escaped or ran aground. They fired their cannon at shore-based semaphore stations and harbors and their young midshipmen and brawling sailors raided military installations to kill soldiers, burn supplies and destroy Napoleon's buildings. It was under the instructions to "cause annoyance" that the British warship, one of the squadron commanded by Collingwood, intercepted Arago's convoy as it approached Marseille; the ships were obviously doing something which would become useful to France, unless prevented.

On Friday 30 June, 1809, as *Minstrel* approached Marseille from the south, the day began as cloudy and rainy with distant rumbles of thunder and lightning. It was a warm summer day, with thunderstorms threatening in the mountains on the shore, but the weather cleared at dawn, and at 7:30 a.m. the lookout at the mast head saw three sailing vessels to the south west heading for Marseille. The *Minstrel* set all sails to chase them and outran them, catching them at 9:20 about 30 km from their destination. The three vessels hoisted "Algerine" colors and stopped sail. Some of the crew from the *Minstrel* boarded them and detained two of the vessels "on suspicion of their carrying on Illicit Trades."

The *Minstrel* put a petty officer and a party of men on board the two detained vessels and escorted them overnight to Collingwood at the Hyères Islands. In the National Maritime Museum, Greenwich is Collingwood's journal (NMM reference COL 13b) in which he records letters he has written to the Admiralty in London and other day-to-day-occurrences. On 1 July, 1809 he records *Minstrel's* success in capturing two Algerian vessels the day before, but there are no other incidents like the one described by Arago in his autobiography. Collingwood's journal says that the sloop HMS *Minstrel* detained "two ships under Algerine colours bound for Marseille from Algiers." He goes on to state, "At noon, the *Tigre* detained a schooner which was passing near the fleet under similar circumstances. Distributed the Algerines to several of the ships and ordered the commander of the *Minstrel* to proceed with the vessels immediately to Gibraltar from whence he was to rejoin me as speedily as possible." With the schooner in tow, *Minstrel* escorted its convoy of two other prizes to Gibraltar, which they reached on 17 July 1809. *Minstrel* returned to the Toulon blockade and on 21 October 1809 took part in a major battle with the French fleet as it tried unsuccessfully to break out from Toulon to run a convoy to Spain to re-supply the army there.

In the letter Collingwood wrote to the British consul in Algiers, Mr H. N. Blanckley (the letter is in the National Archives, TNA reference FO 3/11), he explained the British perspective on the illegality of the trade that Arago was party to:

Some of His Majesty's Ships cruizing before Marseilles have taken several Ships and Vessels having the Flag of Algiers and laden with merchandise the produce of the West Indies and of Spain – consequently these Ships acting in direct opposition to His Majesty's orders in Council which permit the Vessels of the Barbary States to trade freely with the produce of their own Country to ports in France but prohibit the produce of other countries being carried by them – have been detained and sent to Gibraltar … it could not be understood that while the Ports in France were by the decree of Bonaparte shut to all ships of European Powers – that the produce of their colonies should be carried to them from Algiers.

The decree mentioned is the so-called Continental Blockade, by which Bonaparte was attempting to subdue Britain by denying its trade with the continent of Europe which was mostly by then under Napoleon's control. In its blockades of French and Spanish ports (Toulon, Cadiz, Brest, etc. and Marseille), the Royal Navy was attempting to reciprocate by denying France trade with countries outside the Napoleonic Empire.

There is a discrepancy between Arago's statement that there were three merchantmen and the naval records that Minstrel detained two. Perhaps the schooner that Collingwood records as having been detained by *Tigre* was the third. There is also a discrepancy of one day in the dates, but Arago was writing forty years after the events, using diaries that he had been carrying during several years of adventures, whereas Collingwood and the *Minstrel's* officers were recording the events as they happened, keeping records under naval discipline.

Minstrel had let Arago's ship go free. In his autobiography, Arago says that his ship was carrying cotton, which was probably from America, but to Commander Hollinworth the cargo apparently seemed Algerian. The ship was also Algerian, or at least it must have seemed so from the evidence that was produced to Hollinworth when he boarded the ship, since he did not take it as a prize. Arago himself was a French official and might have been taken prisoner if Hollinworth had identified him as such. A young student, he may not have looked like a French official after all his adventures, and presumably (and somewhat uncharacteristically) he had kept his mouth shut thus giving no evidence to sound like one. The crew of the Algerian ship must have all appeared Algerian to Hollinworth and therefore he concluded that the ship was not in violation of the terms of the Order in Council establishing the blockade (although in actuality it was). The ship, including Arago, was let free immediately after its capture on 30 June, 1809.

Arago was lucky not to have been detained. If he had been taken prisoner, he would probably have been sent to Plymouth (via Gibraltar) and been imprisoned at one of the prison hulks in the harbor. He might have been then sent to the gray, granite Prisoner of War Depot that had been newly constructed on a foggy tor on Dartmoor (now Dartmoor Prison) and was the home to several thousand French prisoners. As a French civilian prisoner he might, at best, have been paroled to live in some small Devon town, able to enjoy all the amenities that he could afford and that were available within a mile of the town boundary. (Need it be added that what was available in these towns would not have compared with what was available in the sophistication of Paris?) True, there would have been the female company that he so enjoyed but not the freedom, nor probably the mathematical and other intellectual discussion that, as a student, he would have sorely missed. Certainly the conversation would not have been at his level of capability, and he would probably not have been released until 1814.

After its release, Arago's ship anchored for an overnight stay in the shallow-water natural harbor of the island of Pomègue, one of a group of four in the beautiful Frioul archipelago that lies 5–10 km offshore from Marseille. Another of the islands is famous in literature as the supposed site of the imprisonment in the Château d'If of the Count of Monte Cristo in the novel of the same name by Alexandre Dumas. The Castle looms up massively above the rocky coastline, its inner towers protected under concentric walls and ramparts, and other islands in the archipelago have similar, but smaller, fortifications. The harbor was later used to isolate suspicious ships for inspection (e.g. against the importation of the plague) before they were allowed to dock in Marseille.

Arago says in his autobiography that after releasing his ship, Commander Hollinworth realized he had made a mistake and attempted unsuccessfully to rectify it. The British warship returned to the scene and made moves to capture the vessel that it had freed, putting her long boats to sea over the course of the night and storming Pomègue island and the Castle. Arago says that the British abandoned the attempt to re-capture his ship as too dangerous, presumably because they were repelled by the military garrison. There is no record of this incident in the *Minstrel's* logs, and the incident does not fit with the chronology of events given by Hollinworth, Collingwood and another captain who encountered *Minstrel* during the day. Perhaps Arago misinterpreted or misrepresented other activity by another British warship, HMS *Volontaire*, which sent sailors on a raiding action a few days earlier onto the island of Pomègue (or possibly another nearby island), but the raid would have been being freshly talked about when he arrived on Pomègue. Arago's story of his return to France may have improved with re-telling, incorporating elements of stories he heard on the island and during his subsequent quarantine in Marseille about the raid by the *Volontaire*, until, by the time he re-presented it and wrote it down, the account had reached the printed version and confused the two ships and the two separate actions.

Arago had avoided yet another period of imprisonment. At the time that the *Minstrel* was passing south of Marseille, escorting the Algerian merchantmen westwards into captivity in Gibraltar, Arago once again set foot on French soil at the *lazaretto de Marseille* (the quarantine area). He says this was on July 2, 1809, but in actuality it was probably the day before. Within a few days of arriving Arago stepped ashore in Marseille and wrote that he was relieved to be home "on my native soil" and at "the termination of my laborious and perilous adventures." A voyage from Majorca to Marseille that should have taken a few days had taken nearly a year, and a scientific expedition that should have taken a few months had taken three years.

ARAGO WAS WELCOMED to France by Jean-Louis Pons (1761–1831), the Director of the Observatory in Marseille, who became well known as a discoverer of comets and asteroids. At that time the Observatory was located in the district known as Le Panier, on a hill a short walk from the harbor of Marseille (the

Observatory building is now a school). He was also welcomed by the naturalist Alexander von Humboldt (1769–1859), who was fresh back from his own adventures exploring South America. He had taken up work in Paris with Gay-Lussac, and he invited Arago to share an apartment upon his return to Paris (from 1809–1811). The fact that Arago was feted in this way by such illustrious scientists indicates the respect that he had earned from the scientific community.

When his quarantine expired Arago visited his mother and father in Perpignan. They had feared him dead because his father had seen a captured Spanish soldier in Perpignan with his son's watch. Arago had sold it in Rosas the previous summer for food such as herrings, grapes and lard. Assuming the worst, his mother had had church masses said for his soul and she was emotional at seeing him return, as she thought, from the dead. After this joyful reunion, Arago returned to Paris.

In all his adventures, Arago had safeguarded his precious observations, carrying papers stuffed in the front of his shirt and protecting them as he protected himself. He was thus able to deposit them at the Paris Observatory and when the observations were reduced, he was able to confirm the shape of the Earth by tabulating the length of one degree of latitude across Europe, from London to Formentera. Even over the ten-degree latitude difference from the north to the south it was clear that the degree got longer towards the pole.

Length of one degree on the meridian

Name of arc	Mean latitude	Length of a degree in toises
Greenwich to Dunkerque	51° 15′	57,097.6
Dunkerque to the Pantheon	49° 56′	57,087.7
The Pantheon to Evaux	47° 30′	57,069.3
Evaux to Carcassonne	44° 42′	56,988.4
Carcassonne to Montjuïc	42° 17′	56,960.5
Montjuïc to Formentera	40° 01′	56,955.4

With such convincing observations and a heroic story to tell, Arago's return was a sensation and on 18 September, 1809, at the age of 23, he was nominated to the Academy of Sciences to replace Joseph Jérôme Lalande. His election was not, however, straightforward; Pierre-Simon Laplace had been lobbying to elect another scientist, Siméon-Denis Poisson, at the time of the first available vacancy in the Academy. Siméon-Denis Poisson (1781–1840) was five years older than Arago and when Arago was still a student, Poisson had been his professor at the Polytechnic School. Poisson was not only a gifted mathematician but also had cultivated social connections in the salons of Paris, the world of the theatre and cultural life. He had been taught by Lagrange and Laplace, both of whom had a high opinion of his mathematical ability. Their high opinion was justified by Poisson's eventual mathematical achievements in the theories of gravity, probability and the flow of radiation; modern mathematicians still speak of "Poisson's Equation" and a "Poissonian distribution of random variables."

Laplace thought it unseemly that the less experienced man should be elected first to the Academy. He sought to undermine Arago's scientific achievements, which were at the time and later on, considerably less than those of Poisson (even though Arago boasts of his fine achievements up to then in his autobiography). Arago must have had mixed feelings about competing with Poisson, since it was he who obtained Arago his first scientific position at the Paris Observatory. Regardless, Arago's fame, heroism and potential for later scientific achievement carried the day, and he was elected by a substantial majority (Poisson was elected three years later).

All this must have been a triumph for Arago, and the adventure and its consequences certainly changed his life. He turned his back on the Polytechnic School, terminated his leave of absence and his potential career as an artilleryman, and became a scientist. His subsequent career was less physically adventurous but scientifically and politically more distinguished. He married in 1811, living in an apartment in the Paris Observatory, and had three sons. He became director of the Paris Observatory in 1834 and took up investigations into the wave theory of light sparked by the theories of Thomas Young (1773–1829) and applied it to stellar aberration, the displacement of star positions by the combination of the motion of light waves from a star and the motion of the Earth round the Sun. He suggested the crucial test between the wave theory of light and the particle theory by comparing the speed of light in water and air and is well-known to astronomers as the person who suggested to Urbain Leverrier (1811–1877) that he should investigate why Uranus was departing from its calculated orbit and might be pulled off track by a hitherto unknown planet. This led to the discovery of the planet Neptune.

Arago was an extraordinary public lecturer, pursuing his democratic beliefs in the education of the people, and left a collection of essays on astronomy that became a best-selling set of 4 volumes called *Astronomie Populaire* (*Popular Astronomy*). In 1851 he arranged a scientific display that caused a sensation in Paris known as the Foucault Pendulum and invented by Jean Foucault.

A FOUCAULT PENDULUM is a pendulum that is not fixed to swing in a plane by any mechanical restriction (as in a pendulum clock, where it is constrained within the case) and can choose the plane in which it swings. It tries to swing back and forth along a plane defined relative to the fixed stars (Crease 2003). As the Earth turns in space, the swing of the pendulum rotates, revealing the Earth's motion.

People Jean Bernard Léon Foucault (1819–1868)

Foucault was the son of a publisher and at first trained in the field of medicine. Unable to bear the sight of blood, he turned to physics and with Armand Hippolyte Fizeau he carried out a series of experiments on light and heat. At the suggestion of François Arago in 1850, he established that the speed of light in water was less than in air, a prediction of Young's wave theory of light. His memorial in Montmartre Cemetery records what were thought of as his greatest discoveries: the photoelectric microscope (it had an intense electric arc light source), the eponymous pendulum, a gyroscope, a telescope and regulators.

Inspired by seeing how a vibrating rod of steel clamped in a rotating lathe tended to remain oscillating in the same plane, Foucault became interested in the motion of a pendulum and whether it would stay in the same plane even though the Earth was rotating underneath it. He carried out some preliminary experiments to make a free-swinging pendulum in the basement of his house in the Rue de Vaugirard just to the north of the Paris Observatory at the intersection of the Rue d'Assas (the house is now demolished, but its site is marked with commemorative stonework). The results were encouraging; however, he found that to make the results clearer it was necessary to use a massive pendulum bob less likely to be disturbed by air currents. To see the rotation of the plane of the swing of such a large pendulum bob, it was necessary to get a large swing and therefore a long pendulum. With the cooperation of Arago, he installed one that was suspended from the high ceiling of the Salle Méridienne of the Paris Observatory in February 1851. He confirmed that the slow swing of the long pendulum clearly turned over the floor as a result of the Earth's rotation. Through the Academy, Arago arranged for the experiment to be repeated in public, sending out dramatic invitations inviting all "to come to watch the Earth turn in the central hall of the Observatory."

As a result of all the interest, Foucault set up the experiment under the even higher dome of the Panthéon where it caused a public sensation (Foucault's original pendulum bobs are preserved in the Musée des Arts et Metiers in the Marais on the Right Bank). The Panthéon still has a Foucault exhibition and a replica of the pendulum experiment which still holds the fascination Foucault described a few weeks after his discovery:

> The phenomenon unfolds calmly: it is inevitable, irresistible… Watching it being born and grow, we realize that it is not in the experimenter's power to speed it up or slow it down… Everyone in its presence grows thoughtful and silent for a few seconds, and generally takes away a more pressing and intense feeling of our ceaseless mobility in space.

Few scientific experiments have met with such instant and long-lasting fame and even fewer have been the subject of a novel (*Foucault's Pendulum* by Umberto Eco).

Arago combined a scientific and political career in what continued to be turbulent times. While still a pupil at the Polytechnic School he had shown signs of political principle by refusing to swear allegiance to Napoleon when he seized the crown. He was saved from retaliation by his position as a brilliant mathematician, and Napoleon appears to have been magnanimous in forgiving this youthful idealism; he authorized Arago's appointment as professor at the Polytechnic School soon after his election to the Academy. Arago then visited Britain in 1818 or 1819 with Biot, his mentor at the Paris Observatory, who accompanied him to Spain (see above), to connect the geodetic systems of Britain and France. The pair also visited Unst, the most northern of the Shetland Islands, to measure the period of a pendulum in order to further test the shape of the Earth at high-northerly latitudes. He made a second visit to Britain in 1834 to meet British scientists.

Napoleon abdicated in 1814 and was forced into exile in Elba, where he escaped only to be exiled again in St. Helena. The Bourbon monarchy was restored as the government of France, with the accession of Louis XVIII from 1814–1824 and followed by Charles X from 1824–1830. From 1830 on, Arago was drawn into politics by his younger brothers and was active as a prominent republican. He was a member of the provisional government which took power after the 1848 Revolution and became known as the Second Republic. Sixty years after the Revolution, the political changes had not altered the condition of the people, the socialists declared, and the government was forced to set about a program of reform.

The government established universal manhood suffrage in France and on Arago's initiative abolished slavery in the French colonies. It set up national workshops to guarantee work for everyone by implementing various public projects. Arago was minister of war and marine, and his brother Étienne and son Emmanuel had important posts in the same government (Emmanuel continued his distinguished career in politics long after). Arago voted with the socialists and introduced many reforms while holding this government position, outlawing corporal punishment in the navy and improving the condition of the sailors. Elected as President of the Commission at the head of the Assembly, he was even the Head of State (or one could say he was the 25th Prime Minister of France) for 46 days in 1848, a level of distinction held by few scientists. As President he implemented an unpopular decision to abolish the national workshops in Paris and replace them with conscription into the army. The Commission was overthrown by a series of riots in June and July 1848 and Arago returned to the Observatory.

Arago's dream of establishing a republic in France came to an end in 1851 in the coup that ended the Second Republic. A year later, Louis-Napoleon re-established the hereditary empire and took the title of Napoleon III. Arago was reluctant to accept the reality of the new situation and refused to take the oath of allegiance to Napoleon III. He was allowed to remain director of the Observatory until he retired back to his birthplace in the Pyrenees in 1853.

The observations made in Spain and the Balearic Islands by Biot and Arago were analyzed and presented by Biot in 1810 who also measured the time kept by a pendulum clock at Bordeaux and Dunkirk. In 1817, he went to Scotland and the Shetland Islands to confirm the geodesy carried out by Colonel William Mudge, director of the Ordnance Survey, the most accurate and comprehensive mapping authority in Britain. In 1813, Mudge had extended into Scotland the accurate survey begun in England and Biot located the maps in astronomical terms of latitude and longitude. Biot revisited Spain to repeat the geodesic measurements made by himself and Arago and to conduct pendulum experiments. He began to realize that the Earth was not a simple ellipsoid of revolution, and that the time kept by pendulum clocks on the same latitude varied locally. His most important scientific work was in the polarization of light, and his name is commemorated in the Biot-Savart Law, which describes the magnetic field set up by a steady electric current and also the Biot Number, the ratio of the rate of conduction of heat in a body and the rate at which heat is radiated from its surface.

ARAGO WAS COMMEMORATED as a scientist and as a statesman in Paris in 1893 by the erection of a bronze statue on a pedestal in the Place Île de Sein. The Place lies south of the Observatory (on the meridian) and in the appropriately named Boulevard Arago. The statue was taken down and melted in 1942 during the occupation of Paris in the Second World War, and the pedestal is now empty. His body is buried in the Cemetery of Père-Lachaise in Paris, to the east of the city center, near the entrance in the Bvd de Ménilmontant and is commemorated there by a bust mounted on a pillar (Division 4, to the south of the Avenue Principale between the Bureau de la Conservation and the Monument aux Morts). In addition, a bust by Pierre-Jean David d'Angers (1788–1856) is exhibited in the Louvre museum and a mural of him by A. Sauvage decorates the café-tabac at 53 Bvd Arago, on the corner of the Rue de la Glacière. His is one of the seventy-two names of French scientists and engineers that casted into the side structures of the Eiffel Tower, under the first balcony and on the side opposite the Military Academy. Apart from these there are no public representations of the man in Paris, although there are statues at his birthplace. This is an anomaly; as a result, on the bicentenary of his birth and at the suggestion of the Arago Association, the City of Paris and the Ministry of Culture decided to make up for the absence in modern times of a memorial in Paris to Arago. They set up an open competition for ideas, which was won by Jan Dibbets.

Places The seventy-two scientists and engineers of the Eiffel Tower

Name	Description
Marc Seguin	mechanic
Joseph Jérôme Lefrançais de Lalande	astronomer
Henri Tresca	engineer and mechanic
Jean-Victor Poncelet Gernald	geometer
Jacques Antoine Charles Bresse	civil and hydraulic engineer
Joseph Louis Lagrange	mathematician
Jean-Baptiste-Charles-Joseph Belanger	mathematician
Georges Cuvier	naturalist
Pierre-Simon Laplace	mathematician and astronomer
Pierre Louis Dulong	physicist and chemist
Michel Chasles	geometer
Antoine Lavoisier	chemist
André-Marie Ampere	mathematician and physicist
Micel Eugène Chevreul	chemist
Jeugène Flachat	engineer
Claude-Louis Marie Henri Navier	mathematician
Adrien-Marie Legendre	geometer
Jean-Antoine Chaptal	agronomist and chemist
Jules Célestin Jamin	physicist
Joseph Louis Gay-Lussac	chemist
Hippolyte Fizeau	physicist
Jacques Schneider	industrialist
Henri Louis Le Chatelier	chemist
Pierre Berthier	mineralogist
Jean-Augustin Barral	agronomist, chemist, physicist
Albert De Dion	engineer
Ernest Goüin	engineer and industrialist
Alexandre Louis Jousselin	engineer
Paul Pierre Broca	physician and anthropologist
Antoine Henri Becquerel	physicist
Gaspard-Gustave Coriolis	engineer and scientist
Jean-François Cail	industrialist
Jacques Triger	engineer
Henri Giffard	engineer
François Perrier	geographer and mathematician
Jacques Charles François Sturm	mathematician
Augustin Louis Cauchy	mathematician

(continued)

(continued)

Name	Description
Eugene Belgrand	engineer
Henri Victor Regnault	chemist and physicist
Augustin-Jean Fresnel	physicist
Gaspard de Prony	engineer
Louis Vicat	engineer
Jean-Jacques Ebelmen	chemist
Charles-Augustin de Coulomb	physicist
Louis Poinsot	mathematician
Léon Foucault	physicist
Charles-Eugène Delaunay	astronomer
Jean-Baptiste Morin	mathematician and physicist
René-Just Haüy	mineralogist
Émile Combes	engineer and metallurgist
Luis Jacques Thénard	chemist
Dominique François Jean Arago	astronomer and physicist
Simeon Poisson	mathematician and physicist
Gaspard Monge	geometer
Jules Petiet	engineer
Louis Daguerre	artist and chemist
Charles-Adolphe Wurtz	chemist
Urbain Le Verrier	astronomer
Albert Auguste Perdonnet	engineer
Jean Baptiste Joseph Delambre	astronomer
Etienne-Louis Malus	physicist
Abraham Louis Breguet	mechanic and inventor
Antoine-Rémi Polonceau	engineer
Jean Baptiste André Dumas	chemist
Émile Clapeyron	engineer
Jean-Charles de Borda	mathematician
Jean Baptiste Joseph Fourier	mathematician
Marie François Xavier Bichat	anatomist and physiologist
Jean-Pierre Sauvage	mechanic
Théophile-Jules Pelouze	chemist
Nicolas Léonard Sadi Carnot	engineer
Gabriel Lamé	geometer

Dibbets is a conceptual artist, born in 1941 in Weert in the Netherlands and was well known for his work with photography. His installation sculpture is a homage to Arago and is laid out along the whole of the Paris Meridian inside the Périphérique, (the City's inner ring road), from the Cité Universitaire in the south of the city to Montmartre in the north. It runs across a diameter of the Périphérique, a distance of 8 km, cutting through the following arrondissements: the 14th (Observatoire), the 6th (Luxembourg) on the Left Bank of the River Seine, the 1st (Louvre), 2nd (Bourse), 9th (Opéra), and 18th (Montmartre) on the Right Bank. It seems likely that the monument is the largest dedicated to an individual in France or perhaps the world.

The sculpture consists of 135 bronze medallions 12 cm in diameter which carry Arago's name and the correct orientation N and S. It is interesting that Dibbets chose to mark North and South and emphasized the function of the meridian in pointing north-south, rather than marking them east-west where they would separate the east from the west, or the morning from the afternoon. At first sight the medallions look like the covers implanted in the pavement of access points to utilities like water and gas. The idea was that the people of Paris would accidentally come across the medallions, wonder about them and thus discover Arago.

The medallions are positioned on pavements and paths. If the medallion is in a busy place, the bronze is polished by the passing of many feet. Arago's name is raised in relief and its letters are more polished than the bronze background. By contrast, if the medallion is under trees on a little traffic refuge and out of the way of foot traffic, it is deeply weathered to an olive green. The medallions are accurately located on the meridian (to about a meter) and along the way are other references to the meridian line (Chapter 10).

Chapter 7
Past its Prime

The Paris Observatory (Fig. 6) was the first national observatory and dates from 1667. The Greenwich Observatory (Fig. 34) followed closely (Malin and Stott 1984, Howse 1997), originating with a suggestion by a Frenchman to Louise de Keroulle (1649–1734), Duchess of Portsmouth, who whispered it into the ear of King Charles II. The Frenchman, Le Sieur de St Pierre, about whom nothing certain is known except this reference which originated from the first Astronomer Royal John Flamsteed, claimed that he had solved the answer to the determination of longitude but in truth his suggestion was plagiarized, as I shall explain.

To know where a ship was it was necessary to make observations of the positions of the stars and relate them to calculations of the orientation of the Earth at the time the observations were made. There were no accurate enough clocks, though, so astronomers sought clocks among the stars. Cassini's solution, as we have seen, was to use eclipses of the satellites of Jupiter, but it was not practical to use the long telescopes of the time to observe the satellites of Jupiter from the sailing ship's heaving deck. As an alternative, Le Sieur de St Pierre suggested that the motion of the Moon be used because the position of the Moon could be calculated relative to the fixed stars and you could deduce time if you measured its position,. This was the method of lunar distances first proposed in 1514 by Johann Werner, a German astronomer who edited an edition of *Ptolemy's Geography*, and put forward the principle in a footnote:

> Therefore the geographer goes to one of the given places and from there observes, by means of this observational rod [an instrument that measures angles] at any known moment, the distance between the Moon and one of the fixed stars...

Le Sieur de St Pierre tried to promulgate his idea in France where it received a cool reception, perhaps because the Paris observers were already developing the method which used Jupiter's satellites. He went to England to sell his idea and called on Louise de Keroulle, partly because she was his compatriot and was then the mistress of King Charles II, bearing him a son in 1672. This meeting resulted in King Charles's appointment of a commission to advise him on how to find longitude.

The commission, which included John Flamsteed (1646–1719), agreed that the method was sound in principle, but in practice neither the positions of the stars or the motion of the Moon were known accurately enough nor was there an instrument that could measure to the required accuracy. King Charles decided to attack these problems

P. Murdin, *Full Meridian of Glory*,
DOI: 10.1007/978-0-387-75534-2_7, © Springer Science+Business Media, LLC 2009

Fig. 34 The Royal Observatory at Greenwich. The Greenwich meridian line is marked by a stone and brass construction in the cobbled courtyard beside Flamsteed House, the first observatory on the site, built by Wren. Photo by the author

by appointing Flamsteed as his "astronomical observator," and by founding an observatory to develop the intellectual capital and the technology. He put this project under the charge of the Royal Navy, because of the importance of navigation to its ships. At the suggestion of the astronomer-turned-architect Christopher Wren, the Royal Observatory was built in 1676 by Wren in Greenwich's Royal Park. The description of the person in charge as an astronomical observatory transmuted into Astronomer Royal,

Flamsteed and his successors applied themselves diligently to these measurements and calculations. Flamsteed installed a telescope on a north-south wall with which he made observations of the stars and the planets as they crossed the meridian. From the outset the techniques relied on the meticulous construction of accurate instruments made by scientists and engineers according to continuously developing engineering principles. Flamsteed's successor Edmund Halley (1656–1742) concentrated on observing the motion of the Moon using the theory of gravitation discovered by his friend Isaac Newton in order to develop his Moon motion theory. The third Astronomer Royal James Bradley (1693–1762) installed even larger meridian instruments, and the fourth, fifth and sixth Astronomers Royal (Nathaniel Bliss (1700–1764), Neville Maskelyne (1732–1811) and John Pond (1767–1836) respectively) continued improvements.

It took until the end of the eighteenth century for the methods and measurements to bear fruit. The required data was first published in the *Nautical Almanac* in 1767, under Maskelyne's name. The invention of the required instrument, the octant, was by John Hadley (1682–1744), and was called so because its scale of angular measure was based on one-eighth of a circle. When for practical reasons this was changed to one-sixth of a circle the instrument's name changed to a sextant.

Eventually the method of lunar distances was rendered obsolete by the invention of the chronometer by John Harrison in 1761, and the Greenwich Observatory developed its instruments for determining star positions using transit instruments. Under George Airy (1801–1892), seventh Astronomer Royal, the calculations of the positions of the moon, the planets and the stars were carried out by factory-like methods through human computers or "intelligent drudges" – young men, mathematically well educated but not too much so. Their calculations on forms printed for the purpose were checked at key stages by senior permanent staff. Airy brought the system to a very high state of great reliability, insisting that everything was done in the best possible way (namely *his* way). He treated his staff rather ruthlessly, dismissing the computers when they reached the age of 21 at which time they would be entitled to higher rates of pay.

In 1850 Airy built a new transit circle and this instrument and its observations defined the Greenwich Meridian as it has come to be today, producing observations of unprecedented accuracy. Their accuracy and that of the calculations, coupled with the emergence of Britain as the dominant marine power and the extensive Empire over which it had influence, meant that the British system of latitude and longitude came to predominate over the French, as I shall relate in this Chapter.

HAVING FOUND LATITUDE AND LONGITUDE a sailor would need to relate this to his position on the sea meaning he would need to have accurate maps of the sea, islands and coast lines of the continents. The British maps were executed by the Army using surveying techniques independently of any astronomical measurements and starting on a baseline laid out on Hounslow Heath (now the position of Heathrow Airport[20]). Its northwest end is now marked by an upturned cannon

---

[20] It is not a coincidence that the fundamental bases of both the trigonometrical survey of France and Britain both lie now partly within major airports serving their capital cities (Orly and Heathrow respectively, serving Paris and London). The needs of state-sponsored surveyors for an extensive, accessible and low population-density flat plain were the same as the needs of airlines who needed to land aircraft near their major city destinations.

mounted in the police compound's concrete at the entrance to the Heathrow Airport access tunnel. The precise end of the fundamental base is at the center of the cap screwed over the muzzle. The opposite end is on a similar cap 5.9102 miles away in Hampton, in the Borough of Richmond, 2 km west of Teddington on the road called the A312. The upturned cannon is in Roy Grove, a cul-de-sac off Hanworth Rd. named after Major General William Roy (1726–1790), the initiator of the survey of Britain. The tablet next to the cannon in Roy's Grove states:

> This tablet was affixed in 1926 to commemorate the 200th anniversary of the birth of Major General William Roy, F.R.S., born 4th May 1726 - died 1st July 1790. He conceived the idea of carrying out the triangulation of this country and of constructing a complete and accurate map and thereby laid the foundation of the Ordnance Survey. This gun marks the S.E. terminal of the base which was measured in 1784, under the supervision of General Roy, as part of the operations for determining the relative positions of the Greenwich and Paris Observatories – this measurement was rendered possible by the munificence of H.M. King George III, who inspected the work on 21st August 1784. The base was measured again in 1791 by Captain Mudge as the commencement of the principal triangulation of Great Britain. Length of base - reduced to m.s.l. as measured by Roy 27404.01 feet, as measured by Mudge 27404.24 feet, as determined by Clarke in 1858 in terms of the ordnance survey standard o1, 27406.19 feet.

The military origins of the maps begun by Roy can be seen in the name of the resultant publication, the *Ordnance Survey*. The maps were drawn in their own system of coordinates, the Ordnance Survey Grid. Places that were mapped were linked by a series of surveyor's triangles to Ordnance Survey Triangulation Points strategically located on prominent high points throughout Britain. Unlike the French surveys, the astronomy and the surveying were executed independently and then related later. The Ordnance Survey Grid was linked to astronomical observations by Bradley (the Third Astronomer Royal) at a transit telescope at Greenwich. It remains linked to this Greenwich Meridian today, some 5.7 meters west of the Greenwich Meridian as defined by Airy's transit circle. The separate development of the Ordnance Survey accounts for the rather confusing Grid references and latitude and longitude systems around the edges of the British Ordnance Survey maps of the present time, contrasting the marginal simplicity and scientific elegance of the French maps calibrated simply in longitude and latitude.

Cassini III made several attempts to persuade British sovereigns to carry out surveys on the French model. George III eventually agreed and between 1763 and 1784 the triangulation of England and Ireland was carried out. The surveying instruments were made in England, and, setting aside the French quadrant design, the talented instrument maker Jesse Ramsden developed the theodolite[21] for the survey. A theodolite has two graduated circles at right angles for measuring altitude and azimuth. It was invented in the sixteenth century by Leonard Digges and published as a design by his son Thomas in 1571. It works by a telescope pivoting at the center of each to view the targets. In 1787, Ramsden built a large theodolite with

---

[21] Although its etymology is subject to doubt, the word has nothing to do with God (as in the word "theology") but, apparently the Greek word for "sight" or "view" (as used in the word "theorem").

a 36-inch full circular scale to link the Greenwich Meridian across the English Channel to the Paris Meridian.

This was a well timed project, the American War of Independence having ended in 1783 and the British government keen to show (in a subtle way) reconciliation with France. Of course the project required close intergovernmental cooperation, with the erection of coordinated signal lanterns and the conveyance of messages etc. across the English Channel that lies between the two countries. Given the unreliability of the weather in the Channel, frequent intercommunication was necessary, and it was agreed that both sides would start at the coasts and link to the capitals from there. On the English side, Major General William Roy linked across the Channel from the cliffs above Dover in Kent and near Fairlight in Sussex to church spires and to signal fires lit on the French coast. The Paris Meridian was linked in the other direction across the Channel from Dunkerque to the Pas de Calais and to Dover and Fairlight by the French geodesists including Cassini IV, Pierre Méchain and Adrien-Marie Legendre.

People Adrien-Marie Legendre (1752–1833)

Legendre came from a bourgeois family and showed aptitude in mathematics from an early age. He taught at the École Militaire in Paris and conducted research into mathematics and mathematical astronomy, most especially celestial mechanics (theory of planetary motions). He was elected to the Academy of Sciences in 1783 and as a result of his work in the cross-Channel geodesy as well as a theorem that arose from it on spherical triangles known as "Legendre's theorem," he was made a fellow of the Royal Society. He went on to develop theorems in elliptical functions and the methods of least squares fitting of a curve to data, in which he disputed priority with Gauss. In the Revolution he lost his "small fortune" but re-established his place in scientific life by his brilliant mathematical work.

The difference in longitude of the Paris and Greenwich Meridians was measured by the 1787 expeditions at 2° 19' 51", about 20" less than the current modern value.

Places Fairlight

The area near Fairlight, East Sussex, near Hastings, is now a country park and its highest point is called North's Seat where high spots can be seen up to forty miles away. The nearest point in France is Cap Gris Nez, 61km due east, behind the trees on the skyline near Fairlight Church. In October 1787 William Roy, the military surveyor, triangulated from Fairlight to Wrotham Hill to the north, Dover Castle to the east and across the English Channel to Cap Blanc-Nez and Montlambert in France. He worked at night using light signals mounted on a scaffold 10m high. The heritage of Roy's work lies presently in the bench mark on a concrete pillar although it is not on the original site. Roy's measurement was repeated in July 1825, when Sir John Herschel, the astronomer, and a French officer observed from the same spot some rocket firings in France, timing the flashes in local solar time in order to determine the difference of longitude.

It is curious how in tackling the identical problem of position-finding the British and the French went about it in quite different ways. The French concentrated on the administration of France and the problems of the shape of the Earth, using triangulation and the satellites of Jupiter to determine position. The British were driven by the needs of ships at sea and used the method of lunar distances for the same purpose. In addition, the French were driven by science and adopted an intellectual, integrated approach; the British were driven by practical applications and executed their various programs independently through the application of instrument technology.

Places Dunkerque

The link between the Paris and Greenwich meridians is commemorated at Dunkerque by a monument in the urban park of Fort de Petite-Synthe (off Rue de Nancy running south from the N1), moved here from its original location in the center of the city to a new location on the Paris Meridian. Standing beside a lake, the monument is an obelisk surrounded by four smaller ones. At one time a globe symbolizing the Earth surmounted the obelisk at a height of about 4 meters. The meridian itself runs into the sea in the sand dunes to the north. The fort was originally built in 1878 and re-fortified prior to the First World War; it saw action in the Battle of Flanders in 1940.

JUST AS THE MERIDIANS were linked by astronomy and geodesy, the astronomers who mapped the Greenwich and the Paris Meridians were linked as scientific colleagues, but they were also to become political rivals. The simultaneous existence of scientific cooperation and competition is today dubbed "coopertition." Coopertition over the Paris and Greenwich Meridians came to a head at the Washington Conference in 1884 which was a significant event in the development of the Paris Meridian. In the sixteenth to the nineteenth centuries the Meridian was a living scientific location but today is historic. The change in 1884 rendered it, not obsolete, but certainly less vibrant.

The reason for the Conference was because as trade and global communication became more common in the nineteenth century, it also became necessary to rationalize the systems of longitude and time based on numerous different national observatories. In the second half of the nineteenth century there were nautical maps circulating based on meridians at London (Greenwich), Paris, Cadiz (San Fernando), Naples, Christiania, Hierro (in the Canary Islands), Pulkowa, Stockholm, Lisbon, Copenhagen, and Rio de Janeiro. Land maps were based on these meridians and others at Madrid, Munich, Brussels, Amsterdam, Washington, and Warsaw; it was confusing for everyone. Even as recently as the late 1970's, one of the first telescope buildings erected on the island of La Palma in the Canary Islands at the Observatorio del Roque de los Muchachos was erected at an angle to the cardinal points because of this confusion. The foundations of the building, which now no longer exists, had been laid out by observing the shadows cast by the Sun at local noon and local noon was calculated from a watch reading GMT, applying the correction for longitude of the place. The longitude was read from a Mapa Militár or the Spanish equivalent of the British Ordnance Survey, but not many of these topographical maps were distributed in the Canaries (even in the late 1980's a new printing disappeared from the shops in days) and the one used to orient the telescope building was well-loved and somewhat aged. Unnoticed, its longitude system was based on Madrid. (Spain used the port of Cadiz, more specifically the observatory at San Fernando, for the meridian of the origin of longitude for its navy's maps, but Madrid for its land maps). The longitude of Madrid, 3° 41′W, is not so different from that of Greenwich that the discrepancy in the longitude of La Palma (14° 12′ instead of 17° 53′) was obvious on the ground, so the building was laid out nearly 4° skew.

The problem of the misorientation of the telescope building was inconvenient, but in other cases confusion about longitude might have been dangerous. For example in the nineteenth century it was common for two ships that passed each other to exchange estimates of position for comparison. This might be through large numbers written in chalk on a blackboard, possibly in poor visibility conditions, and with little space or time for geodetic finesse. If there were opportunities for mistakes in the careful laying out of foundations, there would be even greater risks of mistakes in the hurried and limited communications between ships. And what might be the consequences if wrong information influenced a captain in the assessment of his longitude at a later time in the voyage nearer to shore?

The same confusion was possible in analyzing scientific observations established on time. Events such as eclipses or the variability of stars were observed in time systems based on clocks kept at the local observatory, regulated by the passage of the stars. Meteorological and magnetic observations were further examples of sciences where timed observations were required. In order to gather, compare and analyze such observations there had to be a system relating time frames and longitude. This confusion had been noted by Pierre-Simon Laplace in 1800, who advocated the world-wide unification of longitude and by John Herschel in 1828, who advocated unification of time-keeping (Howse 1984).

Some progress was made in unifying timekeeping with the coming of the railways, which involved traveling long distances east-west while keeping to a schedule drawn up accurately to the minute. Alongside the railway lines telegraph wires were erected that carried pulses of electrical current which traveled all but instantly. This made it possible to synchronize clocks at stations along the tracks, and therefore in a small country it became possible to distribute a national time system based on observations at its national observatory; Britain distributed Greenwich Mean Time (GMT) in this way. In larger countries with greater east-west extent such as the United States, the railroads were run by separate owners over limited longitude ranges, and the schedules were coordinated in separate time zones usually based on the railway company's headquarters. In the United States in 1883, there were 49 separate railroad operating time systems. Additionally, submarine telegraph cables made coordination necessary across oceans. The development of transcontinental and intercontinental air travel was still in the future but this would only have added weight to the requirement for a global time system.

THE FIRST OFFICIAL MOVE to unification was made in 1871 at the International Geographical Conference in Antwerp. The Conference resolved that nautical charts should be unified on the Greenwich Meridian and that communication between ships should always provide longitude based this way – but not coastal or land charts; the possession of a meridian had become a matter of national pride. In 1809 William Lambert presented to the United States House of Representatives a memorandum stating that the calculation of longitude from the meridian of a foreign country implied a degrading dependence and was a shackle of colonial dependence which remained an encumbrance unworthy of the freedom and sovereignty of the American people; in essence, the United States ought to have a meridian of its own. It was obvious that if one meridian was chosen as the Prime Meridian, whatever the positive reasons to do so, the others would feel lessened in importance. During the discussion, M. Levasseur, one of the French representatives, generously said that had the discussion taken place in the sixteenth or seventeenth century, the choice of Prime Meridian would naturally have fallen on the Paris Meridian. But the majority of charts being used at that time were based on the Greenwich Meridian and it had become the natural choice.

In 1875, a second International Geographical Conference in Rome re-opened the matter and here France linked time and longitude to the issue of the standardization of units. France would adopt the Greenwich Meridian if Britain would adopt the metric system. Once again, the scientific and practical arguments were linked in political negotiation.

The Americans took up the issue as a civil matter rather than as a scientific one, driven by the practical problem of the range of longitude across the country (Bartky 2000). Charles Dowd of Saratoga Springs, NY, devised the present time zone system, based on strips of longitude 15° wide, with a one hour increment in standard zone time from one to the next. Cleveland Abbe of the Cincinnati Observatory became the first official weather forecaster of the United States and found a uniform time system a pressing need; he was therefore designated as the director of the US Signal Office to do this. In 1876, Sandford Fleming of the Canadian Pacific Railway strongly argued for a unified time that could be used across the world for railway, telegraph and scientific purposes, and advocated the practical use of Dowd's time zone system.

In April 1883, the General Time Convention met in St. Louis to rationalize the US railway time systems. Its secretary, a railway engineer named William F. Allen, showed how to cluster railroad lines into groups to run on standard time zones. He grouped the ten most widely used operating time standards into an eastern and a western set and located the end points of the railway lines in each set. He calculated the central meridian of each pair of extremes and found that they were almost exactly an hour apart in time. He proposed that they be defined exactly as one hour differences and added three more meridians, two in the western USA and one in Canada's maritime provinces thus encompassing the remaining North American railroads. He showed how in practice the one hour time zone shifts would occur predominantly at the end points of the individual railroad systems so that passengers would change time as they changed trains. He noted "a curious fact" that the central meridian of the Eastern Time Zone coincided within six seconds with the seventy-five degree meridian west of Greenwich, five hours from Greenwich time. The American railroad delegates who argued against the use of a standard of time kept at a distant European city were outnumbered by those who did not want the choice to favor one of their US rivals and this removal of inter-city, inter-railroad rivalry won the consensus. By the end of the year and under the commercial imperative to rationalize railroad operation, Railroad Time was implemented on this system, and by December 1884, all North American railroads except two small ones around Pittsburgh had adopted it.

Not every one had the same pressing practical need to integrate longitude and time systems. The British Astronomer Royal, George Airy, rejected the notion that the government should interfere in such a social matter but said that if longitude was unified it should be based on the meridian at Greenwich. The Superintendent of the American Almanac, Simon Newcomb, said that the unification of time was too perfect a plan for the present state of humanity and scorned the idea that other countries should be brought into the scheme: "We don't care for other nations; we can't help them and they can't help us... [I see] no more reason for considering Europe in this matter than for considering the inhabitants of the planet Mars."

These isolationist views did not prevail. As a diplomatic matter, a conference was called in Washington DC in 1884 by the United States to resolve the international confusion and to establish a system of longitude and time zones around the world. The time zones were to be established at 15 degree intervals around the world, one for each hour of a complete rotation of the Earth, and it was necessary to make a choice as to where the time zones would start and where longitude would originate. The USA (and Canada) had made their choice; a system based on Greenwich would be more convenient for the railroad owners.

THE DISCUSSION WAS dominated in Washington in 1884 by the USA, France and Britain (Howse 1984, Harrison 1994). The American delegation was headed by Rear-Admiral Rodgers and supported by Cleveland Abbe. France was represented by Jules Janssen (Fig. 35) of the Paris Observatory, and Britain by John Couch Adams

Fig. 35 Jules Janssen painted by Jean-Jacques Menner. © Observatoire de Paris

Fig. 36 John Couch Adams, as he appeared at the time of the Washington Conference

(Fig. 36) of Cambridge Observatories, both of them distinguished scientists backed up by governmental delegates.

The United States started the Conference in 1884 by tabling a motion that identified Greenwich as the Prime Meridian. France started by saying that it regarded the Conference as one in which the principle would be examined but that a Prime Meridian would not be decided. The United States disagreed, saying that it had called the Conference, in the words of an Act of Congress of 1882 "for the purpose of fixing a meridian proper to be employed as a common zero of longitude and standard of time reckoning throughout the globe." However the Conference agreed to discuss the issues in successive stages, the first on the decision of principle. It proposed that there should be a Prime Meridian and for the USA this was a foregone conclusion. The practical confusion was obvious, as there had been extensive earlier discussion and the recommendations at the conferences in Antwerp and Rome. The Conference agreed unanimously, and having made the decision that they should identify a Prime Meridian, the delegates then discussed which meridian it should be. The United States said that it was not competing to have the Prime Meridian in America, and the Americans proposed the Greenwich Meridian as their practical choice.

People Pierre Jules César Janssen (1824–1907)

Born in Paris, Janssen became the head of the Astrophysical Observatory of Paris at Meudon (1876) and spent his career there studying the solar spectrum. He identified some of the dark lines observed in the solar spectrum as due to water vapor in the Earth's atmosphere. While observing a solar eclipse in India in 1868, he suggested that some of the unidentified spectral lines that could be seen in the spectrum of the Sun's atmosphere when its face was hidden by the Moon were due to a new chemical element (Norman Lockyer simultaneously came to the same conclusion). Their discovery led to the element being named helium (from *helios* the Sun) and was isolated terrestrially in a radioactive ore by William Ramsay in 1895.

People John Couch Adams (1819–1892)

Adams taught himself mathematics at school and there developed a passion for astronomy. He was educated at Cambridge and, when he was only twenty seven, co-discovered the planet Neptune by mathematical calculation. It was known that the planet Uranus was not following its predicted orbit since its discovery in 1781 therefore either Newton's laws of gravitation were wrong or there must be an unknown planet beyond Uranus. Adams worked from known perturbations to deduce the orbit, position and mass of the body that must be attracting it and in 1845 bestowed the position of the new planet to James Challis (1803–1882), director of the Cambridge observatory, who then gave Adams a letter of introduction to the astronomer royal at Greenwich, G. B. Airy (1801–1892). Adams twice tried to call at Greenwich but failed to get an appointment with Airy and was brushed aside. Challis set up a laborious and less than diligent search for a "star" moving near the predicted position. Simultaneously in Paris, Urbain Le Verrier carried out the same investigation as Adams obtaining the same result. Le Verrier sent his prediction to J. G. Galle at the Berlin observatory, where H. C. D'Arrest identified the planet as a new star on a star chart. Adams went on to make two more notable discoveries in astronomy, one in the orbit of the moon and the other the Leonid meteor shower. He was appointed as director of the Cambridge observatory at the age of forty-two in 1861.

The French countered by stating that it was not bound by the Rome conference recommendation and proposed instead that the meridian should be neutral wondering should it be in the Canary Islands (as was the case in classical geography as set out by Ptolemy, whose Prime Meridian was based in the island of Hierro, the most western land that he knew), or the Bering Straits, or at the Great Pyramid (as proposed by the pyramidologist Charles Piazzi Smyth, the director of the Royal Observatory in Edinburgh), or the Temple of Jerusalem? The British delegation replied that the Rome conference had been attended by twelve directors of national observatories and that therefore the scientific opinion was that zero longitude had to be based at an observatory because of the precision required. Given that the observatory had to be in good communication with the rest of the world and a permanent location, preferably under government control, the ideal candidates were Paris, Berlin, Greenwich and Washington. The practical choice was the Greenwich Observatory because this was meridian used by most sea charts and the cost of reprinting these charts with a new meridian on them would be ten million dollars. The Paris Observatory might be relocated because it was near the center of the city, whereas Greenwich was in a park that was distant from the center of London. Little did the British delegate conceive that both the work of both observatories would eventually be relocated, and both would cease to have a scientific function.

Janssen said that he was not pushing the Paris Observatory as the Prime Meridian but argued that it should be neutral and not cut a continent. France had been the first country to conceive and execute geodetic observations to map Europe, America and Africa and therefore it was the scientific principle that should be considered and not the practical use. The intervention of astronomy or geography would lead the discussion away from the object to be attained and economy and custom should not bias the argument. If Greenwich was chosen it would involve France in heavy sacrifices, willing to take on the sacrifices if others showed the same spirit: "If we are approached with offers of self-sacrifice and thus receive proofs of sincere desire for the general good, France has given sufficient proofs of her love of progress to make her cooperation certain."

The French delegation drew attention to the metric system because it was based on the size of the Earth. The delegation commented that "we are still awaiting the honor of seeing the adoption of the metrical system for common use in England." Additionally, Spain said it would vote for Greenwich if Britain and the United States agreed to adopt the metric system. For the USA, Abbe said that the so-called "neutral metric system" was in fact based on the French measurement of the meter, and if the Germans or the British had carried out the measurements they might have proposed something different. In fact Britain and the United States already used the metric system for scientific measurement, and the British delegation joked that it was England that was making a sacrifice by *not* using the metric system.[22]

The argument went back and forth and was essentially going nowhere. The Conference was adjourned for a week and a subcommittee set up to decide on subsidiary issues including the formulation of replies to people who had sent in written evidence. When the Conference reconvened, Sanford Fleming, who had joined the British delegation from the Dominion of Canada, urged the Conference to consider the French proposition as Citizens of the World, not as spokesmen for particular nationalities. If a neutral meridian was chosen it would have to be a new one as all the others were "national." Would everyone accept it or would it simply add to the number of meridians in use? A neutral meridian was excellent in theory but entirely beyond practicality.

Fleming then provided evidence that seems to have been the turning point of the Conference. It was a table showing that two-thirds of the world's ships and three-quarters of its tonnage already used the Greenwich Meridian. Ten percent used the Paris Meridian and Cadiz, Naples and Christiana accounted for about 5% each. Greenwich was the palpable choice but unfortunately it was a national meridian, said Fleming. He proposed that the Prime Meridian should be at 180° to Greenwich, in the Pacific Ocean, lying along a virtually uninhabited line used by navigators as a date line. Longitude would be calculated from here and not expressed as east or west from Greenwich.

---

[22] As we have already seen, Britain adopted the metric system 100 years later, although the British system is still in popular use. It may well pass soon into complete obsolescence as the current generation of schoolchildren grow to maturity, since they are taught the metric system and find the British system used by their parents quaint, rather than traditional and familiar. The French still believe that Britain sticks with its old system because they drink beer in traditional pints when they visit. British Eurosceptics passionately argue against the imposition from Brussels of a European standard, and stick-in-the-mud members of the British Houses of Parliament still argue with nostalgia the merits of the familiar and the traditional. They are all unaware that metricization in Britain has happened.

The Brazilian delegate intervened in the three-way battle between France, Britain and the USA. Brazilian charts were based on three meridians- its marine charts on Paris, its geographic charts on Rio de Janeiro and its telegraphic system on Washington. He would vote for the neutral meridian. Janssen said that it was time to vote and that he had no wish that the discussion should last for ever. In the end the practicalities won out, the least disturbance caused by continuing the de facto situation. The convention adopted the Greenwich Meridian as the world standard by a large majority, with 21 votes in favor and one against. Santo Domingo voted against for reasons that are still unknown to this day; Brazil and France abstained.

The Conference went on to consider the matter of the unified system of time. It adopted a resolution that defined the "universal day" to be the mean solar day starting at midnight on the Prime Meridian, with time counted from midnight in 24 hours; France once again abstained. The resolution was couched in neutral language and did not mention Greenwich Mean Time, but the effect was to define the British national time system as the world standard, under what came to be the "neutral" name of Universal Time.

At the end of the Conference, France proposed a resolution calling for technical studies to regulate and extend the decimalization of angle and time. This followed the proposals of Borda dating back to the Commission on Weights and Measures in 1792. In addition, there was discussion as to whether this was within the scope of the Conference. The conference had been long and tough, people wanted to go home and there was a need to let Janssen walk away from it not entirely as a loser, so the resolution was adopted by 21 votes to none with three abstentions.

The metric unit of angle is the *grad*, a hundredth of a right angle. This unit has some usage in surveying and has been used as a matter of course by the French army (e.g. in setting guns). In the 1970's and 1980's scientific calculators offered the grad as the unit of angle input into calculations of trigonometric functions, as well as radians and degrees, but this convenience for some people seems to have declined in popularity because few people use the metric unit of angle.

The conclusion of the Conference settled the issue of the standard meridian in the minds of the specialists but not in the minds of the legislators. It took many years for the world time system to be accepted by individual countries, and France was certainly not yet ready to yield up the status of the Paris Meridian. In 1895, the sixth International Geographical Congress in London discussed the project of the International Map of the World at one millionth scale. It had been proposed in 1891 by the German geographer Albrecht Penck (1858–1945) who argued that with the exploration of the world virtually complete, the time was ripe to create a map with common conventions of symbols, colors and depiction of topography. France pressed for Paris to be its meridian of zero longitude (Showen 1984) but Britain retaliated by disputing the use of the metric system on the map. This nationalistic dispute was not solved until 1913, when Greenwich and the metric system were adopted for this project. However, the wind was taken out of its sails when the United States seceded from the arrangement in 1921 in despair at the international squabbling and slow progress, and started making its own map of the Americas. The project died a slow death, finally expiring in the 1980s.

The Greenwich Meridian became the world's Prime Meridian by consensus, and the proposals for time zones were adopted pragmatically. France implemented the Greenwich Meridian as the basis for the longitude system of its maps in 1911 and the scientific use of the Paris Meridian faded into history. But like the Paris Meridian, the status of the Greenwich Meridian as an active scientific device has faded as well, as the next chapter (Chapter 8) shows.

Chapter 8
The Greenwich and Paris Meridians in the Space Age

As the viewing conditions worsened in London, meridian observations at the observatory in Greenwich ceased and the instruments relocated to the country village of Herstmonceux in Sussex after the Second World War. The Greenwich meridian did not pass through Herstmonceux and observations made there of stars passing overhead were corrected to the Greenwich Meridian as if made at Greenwich. Even these telescopes were progressively shut down during the second half of the twentieth century as the observing conditions deteriorated even in the countryside in the United Kingdom and better conditions were found on distant, foreign mountains. The Greenwich Meridian was maintained only as a fictitious artifact deduced from telescopes all over the world.

The calculations by which the telescopic observations were combined were carried out at the Bureau International de l'Heure (BIH), established in 1920 at the Paris Observatory as the result of an agreement reached by an international conference in 1912. The Bureau also has the responsibility to combine radio transmissions of time signals into an agreed "exact time" (*l'heure definitive*) which now goes under the name of Universal Time (Coordinated), or UTC. It is an approximation to the time that would be kept by a clock stationed at the Greenwich Meridian. The rivalry in this topic between France and England has persisted into the twentieth century and some English people complained that Greenwich Time and the Greenwich Meridian had ceased to be a fundamental reality located in Britain and had become an abstraction maintained by the French.

Although these developments reduced the reality of the Greenwich Meridian as the prime meridian of the longitude system, the science associated with measurements of terrestrial position improved. The measurements that the large number of telescopes made, and their accuracy when they were combined, increasingly revealed irregularities in the motion of the Earth as it spun on its axis. Both the equator and the North Pole have shifted relative to the continents and the latitude and longitude system has altered accordingly. There are progressive changes in the length of the day caused by the friction of the tides on the land and the post-glacial melting of the icecaps, which slow the rotation of the Earth. There are also seasonal and irregular changes in the rotation of the Earth caused by slippage between its molten core and its mantle as well as by climatic conditions such as the friction of high winds at mountain ranges and the transfer of moisture between the northern

P. Murdin, *Full Meridian of Glory*,
DOI: 10.1007/978-0-387-75534-2_8, © Springer Science+Business Media, LLC 2009

and the southern hemisphere as part of the cycle of summer and winter. As a result, the stars that are seen above the Prime Meridian at any given moment, according to a uniformly beating clock, run either late or early.

The direction of the Earth's axis also changes, wobbling rather like the effect of a sloppy bearing of a rotating machine. Each year the north pole of rotation describes an anticlockwise circle in the Arctic of radius about 10 meters, and the center of the circle drifts westward into Canada at the rate of 15 meters per century. Additionally, continental drift at the rate of millimeters per year moves places relative to the rotation of the Earth. All this means that the system of latitude and longitude, defined by the axis of the rotation of the Earth and the position of the equator, changes and a given place on the Earth adjusts its latitude and longitude with time.

These variations in latitude and longitude started to become very clear when it became possible to compare the positions of places on the Earth and of artificial space satellites. The potential for using satellites to find position became obvious right at the dawn of the space age, with the launch of the first space satellite, Sputnik, in 1957. This satellite carried a beeping radio transmitter, whose sound became famous and was reproduced around the world as proof of the satellite's existence. The beeping could be heard as the satellite orbited across the sky, and its frequency was altered through the Doppler Effect as the satellite approached a radio receiver, passed over-head and then receded. In order to be able to calculate and predict the position and the radio frequency characteristics of the Sputnik satellite, scientists had to input the position of their receiver. It was obvious that you could reverse the problem – if you know the position of the satellite you can calculate where you are.

NAVIGATIONAL SATELLITES are based on the principles described above. Their orbits are determined in space relative to the positions of what used to be called the "fixed stars." The stars are not really "fixed," but move relative to each other as they orbit the Galaxy. Modern astronomers use a better substitute, generally the same idea but more accurately, by using quasars instead of stars. Quasars are exploding galaxies, far away across the Universe and they are certainly not fixed in space but move at high speeds relative to one another. However, any motion they have is diminished by their enormous distance and they form a "fixed" backdrop against which to position the satellites. Think of lying in the grass looking up into the sky. A bee buzzes quickly past your nose and flashes across your field of view, apparently at high speed. But in the same field of view is an airplane, moving much faster than a bee, but at such high altitude that it seems to crawl by comparison with the bee. The airplane is almost fixed. So too are quasars, the positions of which are measured by radio telescopes of a highly sophisticated nature, while the satellites are measured by a global network of tracking stations.

The orbits of the satellites in space are measured accurately from Earth and the orbital information is uploaded into the satellites' computers. It is coded into radio transmissions from the satellites, so that each one tells you where it is and where it is going to be. The satellites also broadcast timing signals, and master clocks kept at the US Naval Observatory control clocks on the satellites which then broadcast signals in unison. The delays as the timing signals travel from the satellites to a receiver determine both the time at the receiver and the position of the receiver relative

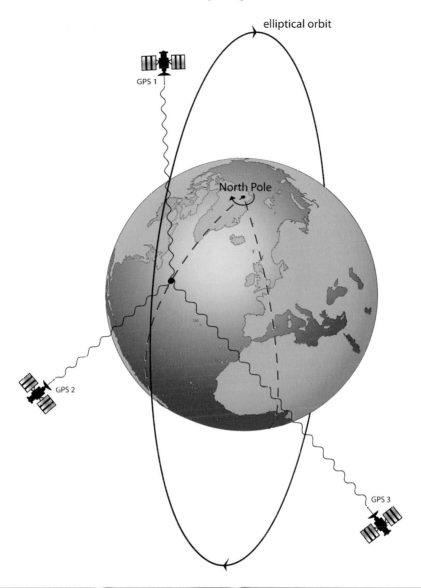

Fig. 37 A satellite navigational system like GPS measures the position of the receiver from the signals sent by three of the swarm of satellites. Time delays between the transmission of the signals and their reception at the receiver provide the location of the receiver relative to the satellites. The receiver calculates the positions of the satellites from a second set of signals that each broadcasts and locates itself on the surface of the Earth, relative to them

to the satellites (Fig. 37). There is more than one system in existence, including one operated by Russia under the name GLONAST, but the system most widely used is a formation of American space satellites called the Global Positioning System (GPS). Stimulated particularly by France as a counter to American domination and in the pursuit of European autonomy (GPS could in principle be turned off at any time by the US military) the European Union is implementing a European equivalent under the name of GALILEO.

In practice, the position of a GPS receiver is measured relative to the tracking stations that determine the orbits of the satellites; the positions of the tracking stations are measured accurately on the Earth's surface, the Earth's surface is measured accurately relative to the center of the Earth, the center of the Earth is located among the satellites, and the satellites and the Earth are located relative to quasars.

It is easily possible to measure the position of a radio receiver to an accuracy of a few meters with the GPS system and even possible to determine the position of one GPS receiver relative to another to millimeters if enough care is taken. The GPS system is used by ships, taxis, delivery drivers, tanks and missiles to navigate to their destinations. Because of the obvious military interest in GPS, the accuracy with which GPS satellites broadcast their orbits to ordinary civilian users is deliberately worsened by its operators (this is called Selective Availability). In its highest form of accuracy, though, it is possible with GPS to measure continental drift and has been used (after considerable scientific effort) to measure how the Indian subcontinent is pushing into Asia and raising the Himalayan Mountains. In time the GALILEO system may be sufficiently easy to use, accurate and robust enough to develop into a means to navigate aircraft through the air from take-off to landing under automatic control.

If the GPS system measures latitude and longitude, however, what latitude and longitude does it measure? Its system has a virtual Greenwich Meridian as a zero point, a result of the exact definitions that it uses for the fundamental parameters of time, position of the Earth, positions of the fixed quasars, etc.; the GPS system is called the World Geodetic System, WGS84 and uses as its prime meridian the one defined by the Bureau International de l'Heure (BIH) at the moment of midnight on New Year's Day 1984. It also uses various conventions about the shape and rotation speed of the Earth to account for its asphericity and wobbles.

This amazing system enables positions on the Earth's surface to be located very accurately and under precise assumptions making it possible to detect changes. In Britain, the latitude and longitude of a given place change relative to WGS84 at the rate of 25 mm per year. Some parts of the world are moving even faster; Australia, for example, moves at 1 meter per decade. In saying what the latitude and longitude of a place are to a high accuracy, it is necessary to say at what date (*epoch*) the coordinates apply.

Independently of the United States military, scientists have agreed to construct a system of 500 reference stations around the world whose positions in WGS84 are accurately measured, including the effects of continental drift. The measurements are combined by BIH into a positional network called the International Terrestrial Reference Frame (ITRF). A sub-section of this is maintained in Europe as the European Terrestrial Reference System (ETRS89). It was identical with WGS84 in 1989, as the name suggests, and has altered since then due to continental drift. It may be thought symbolic that ETSR89 is drifting eastwards relative to WGS84 or that Europe is moving apart from the Americas.

It would be surprising if, after all these changes of definition and improvements in technology, the Prime Meridian remained aligned with the strip laid down at Greenwich and the longitude of Paris remained unchanged from the measurements by the historic French geodesists. The Prime Meridian of ETRS89 in fact differs by 100 meters from the line through the Airy Transit Circle telescope that defines the original Greenwich Meridian at the Royal Greenwich Observatory. All those tourists at Greenwich who get their friends to take photographs as they stand astride the meridian strip have one foot in the western hemisphere and one in the eastern hemisphere, according to the historic definition of the Prime Meridian, but if the tourists used a satellite navigation device to make the journey to Greenwich Park, its GPS receiver will read about 3 arc seconds west of zero. The Prime Meridian according to the GPS system runs across the car park of the Pavilion Tea Rooms and close to the park bench along the ridge footpath that lies beyond the statue of Wolfe. The longitude of Paris has shifted by an equivalent amount. Both the Greenwich and Paris meridians have become in science entirely abstract, and in their modern form are not geographic locations at all.

Chapter 9
On the Trail of *The Da Vinci Code*

Dan Brown's novel, *The Da Vinci Code* (Brown 2003), was first published in 2003. By the end of 2006 its sales were reported to have reached 60 million worldwide, and it had been made into a successful film. Its fast-moving action springs from the murder of an art curator in the Louvre Museum in Paris as the assassin seeks out the location of a powerful religious treasure kept by a secretive organisation. The organisation, the Priory of Sion, is based in Opus Dei, the closed Roman Catholic society, although the secret purposes of Opus Dei in the novel have been much exaggerated and fictionalised. The treasure is called the Holy Grail, but it is not, as in the Arthurian legend, the chalice used by Jesus to offer wine at the Last Supper. It may be the tomb of Mary Magdalene, or it may also be the secret that she is, in the novel, Jesus' wife, a "fact" said to have been hidden by the Priory of Sion for 2000 years. The fifteenth century artist Leonardo Da Vinci, supposedly one of the society's adherents, is reported in the novel to have depicted this secret in a coded way in his painting of the *Last Supper*.

The novel mixes a small spice of fact into a large dollop of fiction to create an entertaining novel of intrigue, adventure, romance, danger and conspiracy, which have been imaginatively worked together to cook up the successful bestseller.

Most interest in the novel's origins has centred on the sensational religious aspects. There are about twenty book-length commentaries on the *The Da Vinci Code*'s religious history. Its account has drawn fire from large calibre guns. In March 2005 the Archbishop of Genoa, Cardinal Tarcisio Bertone, at a seminar called *Storia Senza Storia* ("Story Without History") rebutted the claims that Jesus had a child with Mary Magdalene. He said he wanted "to unmask the lies" so readers could see how "shameful and unfounded" the novel was.

The Cardinal's condemnation may give the words of the novel more significance than they deserve. Dan Brown has written: "All of the art, architecture, secret rituals, secret societies, all of that, is historical fact." This gives an air of authenticity to the novel. Brown has, however, made up the religious doctrines, or based them on questionable accounts by others, or on fiction disguised as fact.[23]

---

[23] The technique is not unparallelled. The Coen brothers' film, *Fargo* (1996), famously opens with the statement "This is a true story. The events depicted in this film took place in Minnesota in 1987." This is a completely fabricated statement to give authenticity to a completely fictional story.

P. Murdin, *Full Meridian of Glory*,
DOI: 10.1007/978-0-387-75534-2_9, © Springer Science+Business Media, LLC 2009

The locations of the actions of *The Da Vinci* Code are not made up. The present book is the scientific story behind the real scenes of several of the novel's actions, those that take place on the Paris meridian. As I have described, this true story is as extraordinary as – possibly more extraordinary than – Dan Brown's fiction. Like the novel, the true story is one of adventure, exploration, danger, attempted murder, murder, foreign travel, political intrigue, religious conspiracy, kidnapping, hostage-taking, escape, romance, love, sex, war, revolution, robbery, lies and deception. The true story is also a story of persistence, dedication, success, achievement, international agreement and honor – and the advance of scientific knowledge.

I am not the only one to have sought out something of what lies behind the *The Da Vinci Code*. Modern-day pilgrims, the novel under arm, can be seen visiting key places in France that are the scenes of action in the novel. There are organised tours of these places. The operators of the Eurostar train that runs between London and Paris said in 2004 that this had been a good year for the number of its rail passengers, mentioning that this was in part due to the effect of the then newly-published novel, driving the curious to the French capital. Likewise, the number of visitors to the Louvre Museum doubled in 2004, the year after the book was published, and increased again to 8.35 million in 2005, when the film of the book was released.

One place of pilgrimage for those interested in *The Da Vinci Code* is the Church of St Sulpice in the St Germain district of Paris, where Pierre Le Monnier built his *meridiana* (Chapter 4). The main entrance to the church is at the top of a steep, wide set of stone steps, above the Place de St. Sulpice. Here, in the book, the albino monk, Silas, gets out of his black Audi and walks up the steps. Greeted by the nun, Sister Sandrine, he says, chillingly, "I prefer to pray alone." In reality he has come to visit the meridiana, though not for its scientific interest, but to search for a clue to the location of the Holy Grail.

Brown adds fictional drama and mysticism to the Church of St Sulpice by saying that it was built over the ruins of a temple dedicated to the Egyptian goddess Isis. This is not so, and a rather huffy notice posted in the Church in 2004 disputed the claim of a "certain recent popular novel" that the Church lies over a pagan temple. Further posters go to some lengths to disprove other details of *The Da Vinci Code*. Certainly no temple walls can be discerned among the Romanesque walls and columns of the 12th century crypt built among the foundations of the Church. But when Silas observes that the Church is stark and cold, almost barren, this gives a good impression of the architecture and layout of its interior.

Silas kneels before the altar at the intersection of the two axes of the cross-shaped plan of the Church, the nave (the longer axis running west-east) and the transept (running north-south). He notes the thin, polished strip of brass glistening in the gray, granite floor, with graduated marks like a ruler. The path of the strip makes its way across the floor, slanting at a rather awkward angle to the transept axis. When it reaches the north wall of the transept, the line is continued vertically along the axis of a specially constructed pillar, or obelisk. It is Egyptian-like in appearance, a marble pillar on a pedestal, narrowing towards the top. I suppose that it was in part the Egyptian appearance that suggested to Dan Brown that there might be a vestige of a pagan temple below the foundations of the Church. Its shape is in fact based on the tapered pillars erected by surveyors in many places in France to mark key measuring locations for the Paris Meridian project. The pillars are pointed because the narrowing to the top well-defines the centre of the pillar as seen through a surveyor's telescope.

In the wall opposite the gnomon there is a high window where the lens of the meridiana was mounted. Nearby is a smaller window containing the letters P and S. In *The Da Vinci Code*, the organization that seeks the Holy Grail of the bloodline of Jesus is the Priory of Sion. In a foreword to the book, Dan Brown says that this "is a real organization" and "a European secret society founded in 1099... In 1975 Paris's Bibliothèque Nationale discovered parchments known as Les Dossiers Secrets, identi-fying numerous members of the Priory of Sion, including Sir Isaac Newton, Botticelli, Victor Hugo, and Leonardo da Vinci." In reality this organization was founded in the middle of the twentieth century and is a hoax. A notice in the church says that "the letters P and S in the small windows at both ends of the transept refer to Peter and Sulpice, the Patron Saint of the church, not an imaginary 'Priory of Sion'."

The brass strip is a meridian line, accurately running north-south and laid out on the floor of the church by Le Monnier. In *The Da Vinci Code,* Silas learns from his Teacher to call the line "the Rose Line". *The Da Vinci Code* gives as an explanation that the name is based on the symbol found on maps used to mark north and, in its full form, representing the 32 directions of the wind – of course north, south, east and west, but also north-east, north-north-east and so on. The simplified version of the symbol is an arrow-head or fleur-de-lis, which marks the north-south direction, but when fully diagrammed inside a circle, the 32 points resemble a traditional thirty-two petal rose flower. The symbol is known as a Compass Rose – hence the name that Dan Brown gives to the meridian line identified by the brass strip in St Sulpice. The name used in this way is an invention of the novel, and is not known to French usage.

The meridian strip in St Sulpice is the location for one of the key scenes of the novel. Silas breaks the tiles at the base of the obelisk to search for a keystone whose markings, he thinks, are the clue to the location of the Holy Grail. In a room above the Church he murders a nun, who has been assigned by a secret society called the Priory of Sion to protect the keystone. In the novel, Silas has been deliberately misled by agents of the Priory of Sion, like the art curator, about the location of the clue to the Holy Grail, in order to protect the secret. However, the hero of the novel, Robert Langdon, a tall, athletic, lean jawed, knowledgeable and intelligent professor (per-haps a model for how the author would like to be…), searching for the hidden treas-ure, finds his way by many adventures to the Palais Royal, just north of the centre of Paris. In the royal arcade he searches out several bronze medallions (Fig. 38), each 12 cm (5 inches) in diameter and marked with the letters *N* and *S*, for *Nord* (north) and *Sud* (south), and uses them as a pointer to the treasure that he seeks.

THESE MEDALLIONS also actually exist. According to reference works there are 135 of them across Paris, running from Montmartre in the north to the Paris Observatory in the south, and beyond to the Cité Universitaire. They mark, says Brown, the "Rose Line" and they cross the courtyard of the Louvre. To suit his purposes to imply historical, religious conspiracy, and just as with the Church of St Sulpice,

Fig. 38 The Arago memorial by Jan Dibbets consists of over 100 bronze disks strewn across Paris along the Meridian, each with Arago's name and its orientation marked

Dan Brown adds a sense of the primitiveness to the Louvre's courtyard by describing it a "once the scene of primeval nature-worshipping festivals". This seems to fit with the extraordinary appearance in the courtyard of the Louvre's glass Pyramid over one of its entrances. I do not know if this bit of the story behind the story is true or not, but the line of medallions does pass near the Pyramid and there is a medallion at its base.

The Pyramid is in fact two pyramids, one above ground, pointing up, one below pointing down into the subterranean mall which lies between the entrance to the museum and its metro station. Brown suggests that the tomb of Mary Magdalene lies below the downward facing point. He identifies François Mitterand, formerly the French President and the person who commissioned the Pyramid as an architectural work, as "a man rumored to move in secret circles". As the novel ends, the conspiracy continues…

The bronze medallions are in reality a memorial to François Arago, whose name appears on each of them, between the letters N and S. His story was told in Chapter 6. In the next chapter, Chapter 10, I describe this memorial across Paris and the journey along it. The journey takes us through the scientific story of this book, past its locations in Paris and scenes of scientific and historic interest.

Chapter 10
Walking the Line: the Arago Memorial

It is 8 km across Paris along the Paris Meridian. It can readily be walked in a day or strolled in a couple of days. Here all the characteristic kinds of neighborhoods of the city are sampled (Fig. 39). The following catalogue lists the locations from south to north of the Arago medallions, according to Benoit Rives on the website of the Paris Observatory, and the number in parentheses is the number of medallions at each site. The number listed is more than a dozen less than the 135 quoted as the number installed and a few of the medallions are located in places to which access is now denied.

I did my field work in 2004 and found that the positions of some of the medallions were obscured by road works and might or might not be uncovered or replaced when the road works are finished. Some I could not find at all, although their locations were clear, because they may have been covered by asphalt; some had been prised off their mounts, perhaps by vandals. The extant number of medallions is considerably less than 135.

This catalogue uses serial numbers to identify the medallions in the pictures and descriptions that follow and to map the route if the meridian is walked from south to north. The meridian line is marked on the commonly used pocket street directory of Paris called *Guide Indicateur des Rues de Paris avec les stations du Métropolitain les plus proches, par Arrondissement et Communes de Banlieue* published by A. Leconte and is the line dividing columns M and N on the maps. Features along the walk are described on the following pages.

> *Take the RER train to the station at Cité Universitaire. Cross the Bvd Jourdan into the campus.*

14th Arrondissement

1–10. Cité Universitaire, on a line between the Cambodian pavilion and the Canadian pavilion, and passing by the Victor Lyon Foundation (10 medallions)

> *Cross Bvd Jourdan into the Parc Montsouris, using the entrance by the RER station.*

P. Murdin, *Full Meridian of Glory*,
DOI: 10.1007/978-0-387-75534-2_10, © Springer Science+Business Media, LLC 2009

Fig. 39 An Arago disk at the feet of a table in a typical Parisian café

11–12. Bvd Jourdan (2)

> *In the Park turn left onto the path that runs parallel to Bvd Jourdan,*
> *and strike across the park at the first medallion past the Meteorological*
> *building to the other side of the park.*

13–21. In the Parc Montsouris on the paths between the Meteorological building
near the Bvd Jourdan and the Cité Universitaire RER station and the
entrance to the park at Av. Reille and Rene Coty (9)

> *Walk along Ave Rene Coty, turn right at Rue de la Tombe Issoire to the Blvd St Jacques and across the Place St Jacques into Rue du Faubourg Saint-Jacques.*

22. Pl. Saint Jacques (1)
23. 81 Rue du Faubourg Saint-Jacques (1)

> *Head towards the Observatory.*

24–29. Bvd Arago/Pl. de l'Île-de-Sein (6)

> *Go into the observatory grounds (Jardin de l'Observatoire). Exit back onto Bvd Arago and circle the observatory grounds via Rue du Faubourg Saint-Jacques and Rue Cassini.*

30–39. Observatoire de Paris: seven on the terrace and in the garden, one in the interior and two in the north courtyard (10)

> *Head along the Av. De l'Observatoire. Off to the left along Bvd de Montparnasse is the Rue Delambre, named after Jean-Baptiste Joseph Delambre (1749-1822) who completed remeasuring the meridian between Dunkerque and Rodez in 1799.*

6th Arrondissement

> *Head along the Av. de l'Observatoire, keeping to the left (west) side.*

40–41. Av. de l'Observatoire (2)
   42. On the median strip at the junction of Av. Denfert Rochereau/Av. de l'Observatoire (1)
43–44. Place Camille Jullian (2)
   45. At the junction between Av. de l'Observatoire/Rue d'Assas (1)

> *Head into the Jardin de Marco-Polo past the Fountain of the Observatory.*

46–48. Jardin de Marco-Polo (3)
   49. At the junction between Av. de l'Observatoire/Rue Michelet (1)

> *Walk along the Av. de l'Observatoire along the left (west) outside edge of Jardin de Robert Cavelier de la Salle.*

50–51. Av. de l'Observatoire, on the pavement on the garden side (2)

> *Enter the Jardin du Luxembourg via the Porte de l'Observatoire (west).*

52. Rue Auguste Comte, at the entrance to the garden (1)

> *Walk by the statue of the alert lion and the child's playground. Keep on the higher level of the garden overlooking the lake and fountain.*

53–62. Jardin du Luxembourg, on the asphalt and concrete areas (10)

> *Detour to the west passing the Museum of the Jardin du Luxembourg. Turn left (westwards) along Rue de Vaugirard to the corner of Rue d'Assas, where stonework on a block of flats commemorates the first pendulum experiment that Léon Foucault carried out on this site. Retrace your steps along the Rue de Vaugirard towards the front of the Senate House.*

63. In front of 28 Rue de Vaugirard, on the side of the Senate House (1)

> *It is convenient at this point to detour to the Panthéon to view the Foucault pendulum set up in the central hall; continue along Rue de Vaugirard, cross the Boulevard St Michel and bear right along Rue Cujas. After the visit, return to the Senate House along Rue de Vaugirard and turn right downhill along the Rue Garanciere to Rue St Sulpice. Turn left towards Place St Sulpice to visit the Church. Retrace your steps to Rue de Seine.*

64–65. Outside 152 and 125–127 Bvd Saint-Germain (2)
66. At the junction of Rue de Seine/Rue des Beaux-Arts(1)
67–70. Rue de Seine, three outside Number 3, one outside Number 12 (4)

> *At the end of Rue de Seine circle around the Institut des Beaux Arts into Quai de Conti.*

71–72. Quai Conti: one at the door of the Institut de France, one in the Institut, near the passage to Rue de Seine (2)
73. Port des Saints-Pères (1)

> *Cross the River Seine by the Pont des Arts, into the Louvre.*

1st Arrondissement

74. Port du Louvre, near to Pont des Arts (1) 75.
    Quai du Louvre, near the entrance to the Pavillon Daru (1)
76–78. Louvre, Denon wing: in the Room of Roman Antiquities, the stairway and the corridor (3)

79–83. Louvre: Napoléon courtyard, east of the Great Pyramid (5)

84–86. Louvre, Richelieu wing, Room of French Sculpture and in front of the escalator to the ground floor (3)

> *Cross the Rue de Rivoli and walk east towards Les Halles (turn left into Rue du Pont Neuf). The headquarters of the French space agency, CNES (Centre Nationale des Études Spatiales), is in Place Maurice-Quentin at the corner of the Les Halles garden. In the Jardin des Halles, above the Forum des Halles, there is a very large sundial designed by the astronomer Dandrel and made by the sculptor Henri de Miller (his more famous sculpture nearby is a monumental stone head in front of the Church of St-Eustache). The sundial, made of fibre optics, is set to show solar time on the Greenwich Meridian and there are extensive instructions on how to interpret it. Return to the Louvre and turn into the Palais Royal, keeping to the left (west).*

87. Rue de Rivoli, entrance to the passage under the arch (1)

88. Pl. du Palais-Royal, on the side of the Rue de Rivoli (1)

> *Exit the gardens of the Palais Royal on the north side, into the Rue de Beaujolais, and turn left into Rue de Montpensier.*

89. At the junction between Pl. Colette/Rue St. Honoré (Conseil d'État) (1)

> *Rue St Honoré in the 18thcentury contained the observatory of the monastery of the Capucins, used by the astronomer Pierre Le Monnier (see the section on the Church of St Sulpice).*

90–96. Palais Royal: Pl. Colette, in the entrance under the arch on the side of the Pl. Colette; in the Nemours Gallery; at the entrance of the passage under the arch leading to the columns of Buren; under the lamp to the left of the Noxa boutique; under the arch in front of the service entrance to the Comédie Française; in front of the cafeteria of the Comédie Française; under the Chartres peristyle, aligned on the black and white tiles between the 4th and 5th line of Buren columns; and in the peristyle of Montpensier at the entrance of the old number 5 rue Montpensier (7)

97. 9 Rue de Montpensier (1)

> *Return to Rue de Beaujolais and go through the passage of the same name into Rue de Richelieu.*

98. 24 Rue de Richelieu (1)

> *Proceed via Rue du 4 Septembre and the Rue de Gramont.*

2nd Arrondissement

   99. 15 Rue Saint Augustin (1)
  100. 16 Rue du 4 Septembre (1)

> *Cross Bvd des Italiens into Rue Taitbout.*

9th Arrondissement

101–2. At the junction between Rue Taitbout /24 Bvd des Italiens (2)

> *Cross Bvd Haussmann.*

103–4. Outside 18/16 and 9/11 Bvd Haussmann (2)

> *Follow Rue Taitbout.*

  104. 34 Rue de Châteaudun on the pavement (1)
105–6. Within the courtyard of the Ministère de l'Éducation Nationale, 34 Rue de Châteaudun (2)

> *At the end of Rue Taitbout turn left on to Rue d'Aumale and right into Rue la Rochefoucauld and on into Rue Pigalle.*

107–8. 69/71 Rue Pigalle (2)
  109. 5 Rue Duperré (1)

> *Turn left into Bvd de Clichy.*

110–1. 21 Bvd de Clichy, on the pavement and in the central median strip (2)

18th Arrondissement

  112. 79 Rue Lepic (1)

> *Continue along Bvd de Clichy and turn right uphill into Rue Lepic. Detour to visit the Place des Abbesses. Retrace your steps along the Rue des Abesses and Rue de Maistre to visit Foucault's tomb in the Cemetery of Montmartre (Divison 7, off the Avenue des Carrières in the northernmost corner of the cemetery).*

> *Above here is the Moulin de la Galette, the location of the northern*
> *sight or "mire" of the Paris Observatory telescopes. Beyond at number*
> *100, on top of the building, there is the dome of a large observatory*
> *constructed in 1860 by David Gruby, and beyond at Number 112 is the*
> *Rue de la Mire, named for the sighting marker for the meridian. Turn*
> *uphill from Rue Lepic at the Moulin de la Galette restaurant and left*
> *again into Av. Junot.*

113. 1 Av. Junot, in the private courtyard (1)
114–5. 3 and 10 Av. Junot (2)
116. 15 rue S. Dereure (1)
117. 45/47 Av. Junot (1)

> *At the Lamarck-Caulaincort métro turn left into Rue Lamarck and right*
> *into Rue Damremont. Turn left into Bvd du Poteau and continue into*
> *Av. de la Porte de Montmartre.*

118. At the junction between Rue René Binet /Av. de la Porte de Montmartre (1)
119. Opposite 18 Av. de la Porte de Montmartre, in front of the entrance to the
      Municipal Library (1)

14th Arrondissement

The southernmost medallions of the Arago monument are in the area of the Cité
Universitaire and Bvd Jourdan, near the Porte d'Orleans area of the Péripherique
ring road. University City was established between the wars as student accommodation to encourage international students to come to Paris and contains hostels in
a mixture of national styles. My wife was there as a summer student and tells me
that, while the sleeping accommodation was segregated, the sexes mixed in the
common room, named by the women students the *salle de chasse* (hunting room),
although who was hunting whom she will not say.

Parc de Montsouris

The Parc de Montsouris (near RER station Cité Universitaire) is a draw for the residents of the area. Young men play football on the flattest area of the grass, young
women walk their dogs past the young men, women and nannies walk their babies,
children run about and graze their knees, and groups of older people sit, gather to
gossip or play chess. The RER railway running to the south of the park and over
bridges to its south do not spoil the charm of the park and its curiosities.

At its northern edge, west of the RER station, the path runs alongside the Bvd Jourdan. Near the meteorological station stands a 12 meter high obelisk with a circular hole at its top. This is a sighting column (*mire*) constructed in 1706 during the rule of Napoleon as a reference point to calibrate the telescope used to define the Paris Meridian. The target reference mechanism was mounted in the metal framework in the circular hole.

DU REGNE
DE
[NAPOLEON I – the name has been removed]
MIRE
DE
L'OBSERVATOIRE
MDCCVI.
(In the reign of Napoleon I, the sight of the Observatory, 1706).

The column is about 50 meters or 5 park benches west of the meridian (Arago medallion Number 13 is mounted in the middle of the pathway). In addition, an observatory operated in the park from the second half of the eighteenth century for the purpose of making observations of interest to the French navy such as positions of navigational stars and the moon, magnetic observations, etc.

Medallions Numbers 14–20 line up across the network of the paths of the park. The northwest exit from the Park is located by a gate into Av. Reille and Av. Rene Coty. The gate-keeper's hut stands inside the park to the right of the exit. In front is Arago medallion Number 21.

The meridian runs northwards towards the Observatory, right of Av. Rene Coty. There are no medallions on this section but the meridian is marked by *le Parking de la Méridienne* and the *Villa de la Méridienne*. Medallion Number 23 is within sight of the Observatory and the plinth for a statue of the astronomer François Arago in Rue du Faubourg St. Jacques. The statue is now gone, removed for scrap during the Second World War.

Observatoire de Paris

The plinth for the Arago statue is in the Pl. Île-de-Sein, crossed by the Boulevard Arago. Two-hundred and fifty meters east along the Bvd Arago are the high blank walls of the Prison de la Santé; outside the prison the guillotine was erected for the public execution of criminals. Further along the Bvd Arago at Number 53 and on the corner of the Rue de la Glacière in the café-tabac, are murals by A. Sauvage of a young and heroic Arago, cloaked, standing with his hand on a terrestrial globe. Across the Pl. Île-de-Sein and beyond the empty plinth of his statue are the gates of the observatory gardens. There is a medallion just outside the gates (Number 29).

Through the gates of the Jardin de l'Observatoire the meridian is marked in stone, running towards the main and original observatory building. North of the stone strip, beyond the railings, there is a number of Arago disks mounted in the garden central to the avenue of lime trees running directly to the south face of the main Observatory building. Here there are enclosures of several auxiliary meridian telescopes in the garden and the private areas visible around.

Usually it is not possible to walk through the Observatory area but guided tours of the Observatory are available by advance appointment. An Arago disk is mounted on the patio on the south side of the building and in the parquet of the Observatory's ground floor. By circling the Observatory grounds to the east you pass the Rue Méchain (named after the astronomer Pierre Méchain, and where the studio of Tamara de Lempicka was situated in the 1930's) and come into Rue Cassini (named after the family of astronomers). The front entrance to the Observatory lies ahead, a proud statue of Urbain Leverrier at the center of its front courtyard. Leverrier discovered the planet Neptune by inferring its existence and position from its effect on the planet Uranus. There are two Arago disks in the courtyard nearby.

The Observatory was established during the reign of King Louis XIV. In 1665-1666, Jean-Baptiste Colbert (1619–1683), an influential minister and at the time the finance minister and secretary of state for the King's court, founded the Academy of Sciences of Paris. A group of its members immediately made a proposal to the King, through Colbert, to establish an observatory in Paris for scientific investigations.

One notable physics experiment carried out at the Observatory was that of Jean-Bernard-Léon Foucault (1819–1868), who with the help of Arago constructed a Foucault pendulum in the 11 meter high Salle Méridienne, or central hall for the Observatory.

6th Arrondissement

Av. de l'Observatoire

In front of the observatory is Arago Medallion Number 40. It is the west of the axis of the Observatory and shows how the present Paris Meridian is offset from Cassini's original (see the section on the Observatory). The meridian continues along the western part of Avenue de l'Observatoire. This wide avenue leads to the open areas of the Jardin de Marco Polo, the Jardin de Robert Cavelier de la Salle (explorer of the Americas) and the Jardin du Luxembourg, which lie one after another between the Observatory and the Senate House. In the distance above the Senate House is Montmartre, with the Cathedral of Sacré Coeur on the hilltop. Left of the Cathedral is the Moulin de la Galette (see below). Near here, now hidden among trees and buildings, is a sighting column used by the astronomers at the Observatory as a calibration point for their telescopes. The axis from the Observatory to this column is the Paris Meridian.

The catalogue lists 34 medallions between here and the River Seine, the remaining ones of the Left Bank. Arago medallion Number 44 lies west of and outside the entrance to the Jardin de Marco Polo, on a triangular traffic island in the Place Camille Jullian.

On the pavement beside the railings to the park is a column erected in 2000 as part of the la Méridienne Verte. The Paris Meridian is mapped on the plaque as going from Dunkerque to Barcelona.

In the Jardin de Marco Polo the meridian lies along the axis of the western avenue of trees, passing to the west of the recently renovated and spectacular Fountain of the Observatory (this is its common name; it is more correctly the *Fountain of the Four Quarters of the World*). It was made in 1873 by the architect Gabriel Davioud (1823–1881), one of Haussmann's callaborators. Under Napoleon III, Baron Georges-Eugène Haussmann (1809–1891) was Prefect of Paris and supervised the reconstruction of its center, setting out the main railway stations and the underpinning infrastructure suitable for a mid-nineteenth century city. He created the city of wide boulevards and open spaces that Paris is today. The sculpture at the center of the Fountain of the Observatory is by Jean-Baptiste Carpeaux (1827–1875); turtles and sea horses surround the base of the plinth, which is girdled by shore creatures. On top of the plinth stand women representing the four continents of Europe, Asia, Africa and the Americas. America is regarded as one continent and Oceania and Antarctica are omitted with the licence of artistic symmetry. The women carry the celestial sphere on their shoulders which then contains a zodiacal band.

Under the trees, closely trimmed along the meridian line, are three tables for table tennis. They are made of concrete and a green agglomerate stone with a metal net for play. They have been placed here by someone with a sense of location; each has the central axis of its "court" aligned along the meridian.

Arago Medallion Number 49, on the corner of the pavement of Rue Michelet between the Jardin de Marco Polo and the Jardin de Robert Cavelier de la Salle, has been pried off its base. So too has Number 50, outside the western entrance to the Jardin de Robert Cavelier de la Salle, opposite to the Faculté de Pharmacie.

Luxembourg

There are 11 medallions in the Jardin de Luxembourg. Number 52 lies in the Rue Auguste Comte outside the Porte de l'Observatoire Ouest. Others are mounted on the pathway by the statue of the alert lion and on the cross path by the children's playground. There are two together at the top of the west steps leading down to the lawn and pond. Two more Green Meridian plaques are mounted on stone balustrades at the corners overlooking the pond with the Medici Fountain and the Senate house in the Palais de Luxembourg. The meridian passes down the west side of the Senate House along a path guarded by a gendarme to stop public access. It is therefore necessary to detour to the west, passing the Museum of the Jardin du Luxembourg. Left (westwards) along Rue de Vaugirard is the Rue d'Assas and stonework on the block of flats on the corner commemorates the first pendulum experiment of Léon Foucault carried out in the cellar of his house on this site. To the right (eastwards) the meridian line crosses the Rue Vaugirard, near where it passes alongside the Senate House.

On the face of the Palais de Luxembourg opposite number 36 Rue de Vaugirard is mounted one of the two last survivors in Paris of the 16 marble meters issued by the revolutionary government between February 1796 and December 1797 as a public reference standard, propagating the meter to practical use. The other is at 13 Place Vendôme, on the right bank outside the Ministry of Justice near the Ritz Hotel. (There is a similar meter standard in Lyon outside the Hotel de Ville. Another is preserved in the Mairie (Town Hall) of the town of Croissy-sur-Seine (78 Les Yvelines); there is a modern replica on the wall of the Rue du Metre at the intersection with the Rue des Ponts, the D321).

Panthéon

It is convenient and interesting at this point in the walk to detour to the Panthéon, east of the Jardin du Luxembourg and on the other side of the Boulevard St. Michel. Originally the building was created by Louis XV as a church dedicated to Ste. Geneviève, the patron saint of Paris. It was completed in 1790 but was then taken over by the revolutionaries as a Temple of Reason and became the secular resting place of France's great men and women – statesmen, generals, admirals, and intellectuals like Voltaire, Pierre and Marie Curie. The huge dome now contains a Foucault pendulum, a replica of the one that Arago helped Léon Foucault to install here in 1851.

St. Germain

Northwest of the Jardin du Luxembourg is the Church of St. Sulpice where the astronomer Pierre Le Monnier installed the *meridiana* (see Chapter 6). The Place de St. Sulpice outside the Church is an area of highly fashionable apartments and cafés. The St. Germain area stretches from here to the River Seine, centered on the Place du St. Germain-des-Prés on the Blvd St. Germain (Medallions Numbers 64–65 on opposite sides of the Blvd St. Germain west of Rue de Seine). This area was the intellectual center of Paris, and the haunt of poets Paul Verlaine and Arthur Rimbaud, philosophers Jean Paul Sartre and Albert Camus, singer Juliette Greco and writer Simone de Beauvoir, as they drank, talked, sang, scribbled their masterpieces on coffee-tables and in garrets, and made love. Jazz clubs concentrate in Rue St. Benoit and art galleries in the Rue de Seine and Rue des Beaux Arts, near the École Nationale Supérieur des Arts, Paris' main school of fine art. Now boutique fashion shops are taking over from the bookshops, but although its character has gentrified, St. Germain still retains a rather cultivated version of its traditional atmosphere.

The meridian runs along the Rue de Seine, crossed by the street market in Rue de Buci. Medallion Number 66 is just off the Rue de Seine in the Rue des Beaux Arts, outside an art gallery.

At the end of the Rue de Seine, on the left bank of the river, is the Institut de France. The building was constructed by Louis XIV's architect, Louis Le Vau (1663–1691). The Institut de France was formed in 1795 by the Revolutionary government from an amalgamation of three Académies of the *ancien régime*, namely the Académie des Sciences, the Académie des Inscriptions et Belles-Lettres (historical documents) and the Académie Française. Later additions included the Académie des Beaux-Arts and the Académie des Sciences Morales et Politiques. It has 325 members who, among other things, take the responsibility to study and define the French language. The Institut is often mocked for its conservatism and purism in doing this, as it erects defences to protect the language against outside change and the impact of modern times. Yet one can only admire some of its inventions, such as the charming word *balladeur* as a translation for a "walkman." A *balladeur* was originally a wandering minstrel, and "walkman" is an exciting but rather ungracious trademark. Medallion number 72 is on the Quai de Conti.

1st Arrondissement

The meridian crosses the Seine along the line of the Pont des Arts footbridge.

The Louvre

The Louvre was founded as a royal palace in the 12th century and enlarged by a succession of kings to its present grand form as a renaissance palace, which dates back to the 1540's. The Louvre is built around a magnificent open courtyard, the Court Carrée, and looks outwards to the Jardin des Tuileries. At the center of the courtyard is the artistically controversial modern glass pyramid commissioned from the architect I.M. Pei by former French president François Mitterrand. The meridian cuts across the palace and the courtyard and almost through the glass pyramid itself. The Louvre houses possibly the finest art collection in the world, including three Arago medallions which traverse the museum's Denon Wing (in the Roman Antiquities section, on a staircase and in a corridor). Five others run across the Cour Carrée behind the glass pyramid.

There are representations of several astronomers in the Louvre collection. Pierre-Jean David's bust of *François Arago* is in the Richelieu Wing, in the Puget Court. Jan Vermeer's *L'Astronome* (Richelieu Wing, Second floor, Room 38) is a fictional and rather fanciful astronomer who refers to a globe of the constellations while studying an open book. In addition, *Nicolas Kratzer* in Hans Holbein's portrait was a real person, as can be seen by his cool gaze. He is shown with some of the sundials that he invented or made (Richelieu Wing, Second floor, Room 8). One of the sundials is identical to one that Holbein inserted into his portrait of *The Ambassadors*, which hangs in London's National Gallery (Holbein presumably put it into the picture, along with other instruments, papers, music and art works to flatter the subjects of the portrait by implying how cultured they were). There are also some fine Islamic celestial spheres and other scientific instruments in the Richelieu Wing, Entresol, Room 5.

In Dan Brown's book *The Da Vinci Code,* the hero feels the Arago disks draw him to the pyramid. Dan Brown says that it has 666 glass panes (a number symbolic of Satan), but this is untrue. Below in the subterranean Carrousel shopping area is a second pyramid, inverted and hanging from the ceiling. On the floor below its tip stands a third pyramid, small and made of stone. The tips of the pyramids point at each other and are said in the novel to mark a hidden vault of documents as well as the tomb of Mary Magdalene. It is hinted that MM Mitterand and Pei were co-conspirators in the secret marking of these religious treasures for those in the know. All this seems unlikely, but it is a fact, however, that the Pyramid marks the Paris Meridian which continues to the north.

Palais Royal

Across the Rue de Rivoli, running alongside the Louvre, is the Palais Royal. The meridian continues across the Rue de Rivoli clipping the corner of the Palais Royal. Medallion Number 90 lies under the colonnade to the west of the Palais.

The Palais Royal was built in 1784 and based on St. Mark's Square in Venice. At the outset its arcades were designated for luxury shopping and it was (and remains) a very fashionable place. It housed the mansion of Cardinal Richelieu, houses the theatre of the Comédie Française and at the Rue de Rivoli's end is an installation sculpture known as the Columns of Buren made of black and white marble and slate pillars spread over the southern end of the courtyard. Its sculptor was Daniel Buren (1938-) its colonnades house fashionable shops and restaurants; at the present time it is a treat to take lunch under a shady umbrella of a restaurant of the Palais Royal, listening to a passing accordion player and watching lunchtime parties of office workers playing *boules* on the gravely paths and open areas.

Originally its security guards were the King's Swiss Guards, who were instructed to refuse entry to "soldiers, domestics, or persons who wore caps or jackets, students, street urchins, beggars, dogs and artisans." This strict dress code of former times and its peaceful present-day elegance gives no hint about its former night-time decadence, when it was a place to meet, drink, flirt, quarrel, gamble, fornicate, and fight. The 18th century poet Jacques Delille, known as Abbot Delille, wrote:

> *Fields, meadows, woods, and flowers*
> *Do not grace this garden.*
> *But while we are there and do wrong*
> *We can also put our watches right.*

This refers to a meridian curiosity standing in the south of the lawns of the garden-a so-called "midday cannon" in the tradition of cannons that were fired from observatories across harbors to signal the time to ships at anchor. The cannon at Cape Town, South Africa is still operated in this way, with a tug of a lanyard. Originally in 1786, the cannon at the Palais Royal was similar but it was replaced in 1795 by M. de Chabrol, the Prefect of the Seine, by a small cannon about 20 cm long and mounted on a granite column. It is an inelegant, squat column which had to be strong to withstand the daily kickback of the cannon. Until 1914 the instrument was automatic, in an arrangement invented by an engineer by name of Rousseau to replace a very accurate sundial in the Rue des Bons-Enfants. The cannon was charged with gun-powder and the firing pan positioned under a magnifying glass. On a sunny day, when the Sun transited the meridian and the lens concentrated solar heat, the cannon fired at 12 noon, local time, plus or minus the so-called "equation of time" representing such effects as the eccentricity of the Earth's orbit round the Sun. Since this correction ranges up to fifteen minutes and some of the circumstances of the firing pan are accidental (distribution of powder, intensity of sunlight, etc.), this was not a very accurate method of signalling time but indeed has the advantage of being loud.

The lens mechanism is missing from the cannon in the Palais Royal, and in the 1990's the cannon was triggered at noon by a park attendant with a watch and lit match in hand; alas, this exciting custom has been suspended.

2nd and 9th Arrondissements

Opéra

The meridian cuts across the 2nd Arrondissement, which is one of the smallest and passes through rather anonymous streets. Medallion Number 99 is in Rue St. Augustin.

The Arago medallions continue northwards into the 9th Arrondissement, east of the busy junction in front of the massive Opera House. The meridian passes through a commercial area of elegant hotels, restaurants and busy department stores (Les Galeries Lafayette are within sight). A *flâneur* (timewaster) seated at the café at 24 Bvd des Italiens may scuff his feet on Medallion Number 101 (Fig. 39), with Number 102 around the corner in Rue Taibout. Numbers 103 and 104 are opposite each other near the Hotel Ambassador on Bvd Haussmann.

18th Arrondissement

Montmartre

Today Montmartre has a varied reputation, with sophisticated nightclubs, shows and sordid sex shops concentrated in the area around the Place Pigalle. Medallion Number 111 lies nearby on the Bvd de Clichy but is often walked over by tourists and by touts for the bars. If you accepted their invitations, you could sip a drink of your choice while chatting to bar-hostesses who will quickly and repeatedly quaff drinks of theirs, and promise more than is usually delivered (indeed you would find the bill for their "champagne" rather high). There were road works in the boulevard in 2004 that rendered unproductive my search for the other medallion located at this point.

Above the Bvd de Clichy rises the hill of the Butte de Montmartre. The Rue Lepic winds up its steep sides, leading to the crowded streets around the Place du Tertre. Here artists purvey to the tourists daubed pictures of Montmartre of dubious merit, although they serve well enough as souvenirs. The sex industry and the tourist industry are all mixed up in Montmartre with a multilingual, lively community of long-term residents and transient students as well as hangers-on. The Rue Lepic changes in character quickly, as it rises from Blvd de Clichy, from touristy to village-like. In its mid section it is lined with butchers, bakers, pharmacies, grocers, newsagents, restaurants, and all the small businesses typical of a French neighborhood, including the bar which was the principal location for the French film *Amélie*. Half way along Rue Lepic, the Rue des Abbesses leads back towards the meridian to the Place des Abbesses, where there is an original art nouveau frontage to the Abbesses Métro station by architect Hector Guimard.

Children play here in the evening, perhaps treated to a ride on the merry-go-round while their minders wait for their parents to come home from work. This attractive ambience stems from the nineteenth century concentration of artists in Montmartre and the bohemian life they brought to the area including Corot, Gericault, Renoir, Degas, Cezanne, Braque, and Picasso. The archetypical images of Montmartre from this epoch are those by Toulouse Lautrec of the Moulin Rouge cabaret and its flamboyant dance, the can-can. *Moulin Rouge* is a red windmill, once real and presently a design of neon lights. The cabaret at the Moulin Rouge is now a rather high-priced theater show pitched to rich, international tourists who sip their compulsory drink at little tables and watch the brightly- (but sparsely-) costumed show-girls, and magicians, jugglers, and other variety acts whose appreciation needs knowledge of no particular language.

Back along the Rue des Abbesses and along the Rue de Maistre is Monmartre Cemetery where Léon Foucault is buried, his grave marked by a pillar recording some of his significant discoveries: the photoelectric microscope, his pendulum, a gyroscope, a telescope, and regulators. Other graves in the crowded but quirky and whimsical cemetery are of the scientist Ampère, the novelist Stendahl, the composer Berlioz, the painter Degas, and the film director Truffaut, as well as a tomb with a gold saxophone; the grave of Adolphe Sax, the inventor of the instrument.

The cemetery was founded here outside the city for health reasons. In the seventeenth, eighteenth centuries, and even into the nineteenth century, Montmartre was an isolated village on a hill to the north of Paris. Because of its elevation, a number (up to 30) of windmills were built there but now few survive, and most of the windmills seen there today are pictorial or ornamental replicas. The Moulin de la Galette is the oldest windmill in Montmartre and two of them exist. One is at the corner of the Rue Lepic and Rue Giradon. It is small, below the top of the hill out of the wind, and anyway it is a restaurant, therefore it cannot be the real Moulin de la Galette.

Just below it is Medallion Number 112 in the upper part of Rue Lepic, where it is composed of the business premises of small specialty businesses (lawyers, accountants, antique dealers, and consultancies).

The real Moulin de la Galette is in the park above here, the highest point on the Paris Meridian within the Péripherique. As such, the windmill can readily be seen from the Observatory located left (west) of the Cathedral of the Sacré Coeur. On private land near the windmill is the northern equivalent of the *mire* at Parc Montsouris. The stone pillar is on private land and can be visited only by application to the caretaker at 1 Boulevard Junot. Its situation among suburban clutter is a disgrace to a historic monument.

Medallion Number 113 lies under the Moulin de la Galette, in Av. Junot. Northwards the meridian passes through residential apartments to the Porte de Clignacourt area of the Périphérique ring road (Numbers. 114–119).

Bibliography

References to works cited in the text are made via the Harvard system of author's name and date of publication, thus: (Adams 2005). If not otherwise referenced, quotations are copied from the summarizing and general works that are referenced near the beginning of the chapter or section. I received James Lequeux' new biography of François Arago (Lequeux 2008) too late to be able to take advantage of its comprehensive scholarship.

Adams, M. (2005) *Admiral Collingwood: Nelson's own Hero* London: Weidenfeld and Nicholson.

Alder, K. (2002) *The Measure of All Things: the Seven Year Odyssey that Transformed the World* London: Little Brown.

Anon (1854) "François Arago his life and discoveries" *North British Review*, XX, 459, 1854 – review essay based on "François Arago" by J.A. Barral, Paris 1853, "Discours prononcé au funeraille de M. Arago, le Mercredi 5 Oct 1853" by M. Flourens, and "François Arago" by M. De la Rive, in *Bull. Univ. de Genève*, Oct 1853, **xxiv**, 265; also "Obituary of François Arago" *Monthly Notices of the Royal Astronomical Society*, **14**, 102, 1854.

Arago, F. (1857) "The story of my youth" in *Biographies of Distinguished Men*, translated by W.H. Smyth, the Rev. Baden Powell and Robert Grant London: Longman, Brown, Green, Longmans and Roberts. See electronic text at http://www.gutenberg.org/etext/16775. The French version of "Histoire de ma Jeunesse" is available through http://gallica.bnf.fr/ (Barral, J.-A., ed. (1854–1862) *Arago, F. (1786–1853) Oeuvres completes*, Tome 1, p. 1). See also "Mesure de la Méridienne de France", *ibid*. Tome 11, p. 54.

Bartky, Ian R. (2000) *Selling the True Time: Nineteenth-Century Timekeeping in America* Stanford: Stanford University Press.

Blackett, P.M.S. (1967) "The Lunar Society of Birmingham" *University of Birmingham Historical Journal* **XI** (1), 1, 1967.

Bodanis, D (2006) *Passionate Minds* London: Little Brown.

Brown, D. (2003) *The Da Vinci Code* London: Bantam Press.

Cassini, Jacques (Cassini II) (1720) *De la Grandeur et de la Figure de la Terre* Paris: Imprimerie Royale.

Cassini de Thury, César-François (Cassini III) (1744) *La Méridienne de l'Observatoire Royal de Paris, Verifiée dans toute l'étendue du Royaume par les nouvelles Observations.* Paris. Libraries Hippolyte-Louis Guerin et Jacques Guerin.

Chandrasekhar, S (1995) *Newton's Principia for the Common Reader* Chicago: University of Chicago Press.

Chapman, A. (1995, 2nd edn.) *Dividing the Circle* Chichester: Wiley and Praxis.

Cohen, I.B. (1970) Biography of Delambre in *Dictionary of Scientific Biography* New York: Charles Scribner & Sons.

Cohen, I.B. and Whitman, A. (1999) *The Principia: a new translation* Berkeley: University of California Press.

Cotardière, Philippe de la (2003) *Guide de l'Astronomie en France* Paris: Guides Savants de Belin.

Crease, R.P. (2003) *The Prism and the Pendulum: the Ten Most Beautiful Experiments in Space* New York: Random House.

Daumas, M. (1943, reprinted 1987) *Arago, La Jeunesse de la Science* Paris: Belin.

Daumas, M. (1972, reprinted 1989) *Scientific Instruments of the Seventeenth and Eighteenth Centuries and their Makers* (trans. by M. Holbrook) London: Portman Books.

Debarth, S. (1984) *L'Observatoire de Paris: son Histoire (1667–1963)* Paris: Observatoire de Paris.

Delambre, M. (1827) *Histoire de l'Astronomie au Dix-Huitième Siècle* Paris: Bachelier.

Fernández-Armesto, F. (2006) *Pathfinders : a Global History of Exploration* Oxford : University Press.

Gillispie, C.G. (ed.) (1973) *Dictionary of Scientific Biography* NY: Charles Scribner's Sons.

Gotteland, A. and Camus, G. (1997) *Cadrans Solaires de Paris* Paris: CNRS Editions.

Greenberg, J. (1995) *The Problem of the Earth's Shape from Newton to Clairaut* Cambridge: University Press.

Guedj, Denis (2001) *The Measure of the World* (trans. by Arthur Goldhammer from the French edition, *La Mesure du monde*, Éditions Seghers Paris 1987) Chicago: University of Chicago Press.

Harrison, H.M. (1994) *Voyager in Time and Space: the life of John Couch Adams,* Sussex: The Book Guild.

Heilbron, J.L. (2001) *The Sun in the Church: Cathedrals as Solar Observatories* Cambridge, MA: Harvard University Press.

Howard-Duff, I (1984) "Paris Observatory in 1784 – Cassini de Thury's legacy to Jean-Dominique Cassini" *J. Brit. Astr. Assoc.,* **95**, 1.

Howard-Duff, I. (1986) "D.F.J. Arago 1786–1853" *J. Brit. Astr. Assoc.,* **97**, 1.

Howse, D. (1984) "1884 and longitude zero" *Vistas in Astronomy* **28**, 11.

Howse, D. (1997) *Greenwich Time and the Longitude* London: Philip Wilson and the National Maritime Museum.

Kiner, Aline (2005) « Des églises devenues observatoires » *Sciences et Avenir* Avril 2005, page 82–97.

Konvitz, J. W (1987) *Cartography in France 1660–1848* Chicago, IL: University Press.

Lequeux, J. (2008) *François Arago un savant généreux - Physique et astronomie au XIXe siècle* Paris: Coédition EDP Sciences, Observatoire de Paris.

Malin, S. and Stott, C. (1984) *The Greenwich Meridian*Southampton: Ordnance Survey.

Maupertuis, P. (1738) *The Figure of the Earth, Determined from Observations made by Order of the French King at the Polar Circle* London: Cox, Davis, Knapton and Millar. See also another English translation in Pinkerton (1808), p. 231.

Méchain, P. and Delambre, J.-B. (1806–10) *Base du Système Mètrique Décimal, ou Mesure de l'Arc du Méridien.* Paris: Baudouin, Imprimeur de l'Institut National.

Murdin, P. (2006) "Laborious and perilous adventures – François Arago's triumphant return to France" *Journal for Maritime Research,* 15 June 2006. http://www.jmr.nmm.ac.uk/

Outhier, R. (1744) *Journal of a Voyage to the North in the Years 1736 and 1737.* See translation in Pinkerton (1808), p. 259.

Pinkerton, J. (1808) *A General Collection of the Best and Most Interesting Voyages and Travels in All Parts of the World* (Vol. 1). London: Longman, Hurst, Rees and Orme.

Regional Library of Lapland and the Pello Municipal Library (2006) "The degree measurements in the Tornionlaasko Valley 1736–1737" http://www4.rovaniemi.fi/lapinkavijat/maupertuis/index_eng.html

Rougé, Michel (2006) *Le Gnomon de l'Eglise Saint-Sulpice* Paris: Paroisse Saint-Sulpice.

Showen, R.L. (1984) "Pride and chauvinism in science" *Vistas in Astronomy* 28, 311, 1984.

Simaan, A. (2001) *La Science au Péril de sa Vie* Paris: Vuibert/Adapt.

Smith, J.R. (1986) *From Plane to Spheroid: determining the Figure of the Earth from 3000 B.C. to the 18th century Lapland and Peruvian Expeditions* Rancho Cordova, CA: Landmark Enterprises.

Sobel, D. (1999) *Longitude* London: Fourth Estate.

Tackett, T. (2003) *When the King Took Flight* Cambridge, MA: Harvard UP.

Terrall, M. (2002) *The Man who Flattened the Earth: Maupertuis and the Sciences in the Enlightenment* Chicago: University of Chicago Press.

Terrien, L. (2000) *Saint-Sulpice* Paris: Church of St Sulpice.

Tobin, W. (2005) *The Life and Science of Léon Foucault* Cambridge: Cambridge University Press.

Todhunter, I. (1873, reprinted 1962) *A History of the Mathematical theories of Attraction and the Figure of the Earth* NY: Dover Publications.

Turner, A. J. (1989) *From Pleasure and Profit to Science and Security – Étienne Lenoir and the Transformation of Precision Instrument-Making in France 1760–1830* Cambridge: Whipple Museum of the History of Science.

Van Helden, A. (1996) "Longitude and the satellites of Jupiter" in *The Quest for Longitude*, ed. W.J.H. Andrewes, Cambridge, MA: Harvard University Press.

Whitaker, R. (2004) *The Mapmaker's Wife* London: Doubleday.

Wolf, C. (1902) *Histoire de l'Observatoire de Paris de sa Fondation à 1793* Paris: Gauthiers-Villars.

Acknowledgements

I am grateful to several people for their encouragement and help to be able to tell this story. They include Harry Blom (Springer), James Caplan (Observatoire de Marseille), Allan Chapman (Oxford), Jessica Fricchione (Springer), Mark Hurm (Institute of Astronomy, Cambridge), Peter Hingley (Royal Astronomical Society), Karel van der Hucht (International Astronomical Union, Paris), Odile Dubois (University of Reims), Eliane Jaouipylypiw (Roissy-en-Brie), Margarette Lincoln (National Maritime Museum), Tom Matheson (New Jersey), Dave Minett (Hastings), Amanda Smith (Institute of Astronomy, Cambridge), Curtis Wilson (SJCA). I am grateful to my wife Lesley for her support in researching this book, for translating Maupertuis' "Poem for Christine", and for accompanying me on our own adventures to places described in this book.

Index

Printed in the United States of America